SHORT MESSAGE SERVICE (SMS)

SHORT MESSAGE SERVICE (SMS)

THE CREATION OF PERSONAL GLOBAL TEXT MESSAGING

Friedhelm Hillebrand (Editor)
Hillebrand & Partners, Germany

Finn Trosby
Telenor, Norway

Kevin Holley
Telefónica Europe, UK

Ian Harris
Research In Motion, Canada

A John Wiley and Sons, Ltd., Publication

Library of Congress Cataloging-in-Publication Data

Harris, Ian, 1962-
   Short message service (SMS) : the creation of global personal text messaging / Friedhelm
Hillebrand, editor ; Ian Harris, Kevin Holley, Finn Trosby.
      p. cm.
   Includes bibliographical references and index.
   ISBN 978-0-470-68865-6 (cloth)
   1. Text messages (Telephone systems) 2.   Instant messaging.   I.   Hillebrand, Friedhelm.
II.   Holley, Kevin, 1963- III. Trosby, Finn. IV. Title.
   TK5105.73.H375 2010
   006.7--dc22

                                                                            2009042563

British Library Cataloguing in Publication Data

A catalogue record for this book is available from the British Library

ISBN 978-0-470-68865-6 (H/B)

Contents

Introduction   **xiii**

**1   Communication Networks in the Early 1980s and the Portfolio of GSM
      Services   1**
      F. Hillebrand

1.1   Station-to-station Morse Telegraphy, the Origin of All Modern Technical Text
      Communication   1
1.2   Network-based Communication Services in the Early 1980s   1
      1.2.1   Telephony   2
      1.2.2   Telex, the Forefather of Modern Text Communication   2
      1.2.3   The Advent of Many Faster Transport Techniques for Text and Data   3
      1.2.4   Creation of New, Fully Standardised Text Communication Services   4
      *1.2.5   Creation of Value-added Text Communication Services Based
                on Servers*   5
      1.2.6   The Development of Private Mobile Radio Networks   6
      *1.2.7   Internet, Web Browsing and Email as the Winners in Communication
                in Fixed Networks*   6
1.3   Services Portfolio of GSM   7
      1.3.1   Way of Working in GSM Standardisation   7
      1.3.2   Service Philosophy of GSM Developed from 1982 to 1984   8
      1.3.3   GSM's Fixed-network-service Companions   9
                1.3.3.1   Reference Model   9
                1.3.3.2   Teleservices   10
                1.3.3.3   Bearer Services   11
1.4   GSM Mobile Telephony and SMS – the Most Successful Telecommunication
      Services   12

2   Who Invented SMS?   15
      F. Hillebrand

2.1   Introduction   15
2.2   Clarification of the Terms 'Invention' and 'Innovation'   15
      2.2.1   Invention   15

2.2.2   *Innovation*   16
2.3   Was SMS Invented during the ISDN Work?   16
2.4   Was SMS Invented by Test Engineers, Students or in a Pizzeria Session?   17
       2.4.1   *Was SMS Discovered or Invented by Test Engineers or Students?*   17
       2.4.2   *Was SMS Invented in a Copenhagen Pizzeria Session in 1982?*   17
2.5   A Clarifying Discussion within the GSM Community in Spring 2009   18
2.6   Timetables of SMS Genesis   19
       2.6.1   *Concept Development and Standardisation of SMS*   19
       2.6.2   *Development of SMS in the Market*   21

3   **The Creation of the SMS Concept from Mid-1984 to Early 1987**   **23**
       F. Hillebrand

3.1   The Birth of the SMS Concept in the French and German Network Operators   23
       3.1.1   *Introduction*   23
       3.1.2   *Documentary Evidence that Survived*   24
                3.1.2.1   Roots of SMS in 'Enhanced' Paging Integrated in a
                           Multiservice Mobile Communication System   24
                3.1.2.2   A Surprising Proposal in the S900 Interim System Context   25
                3.1.2.3   The Mutation of an Enhanced Paging Service to a
                           General Short Messaging Service in the Specification
                           Process of the DF900 Trial Systems for GSM   26
                3.1.2.4   A Technical Discussion and Input to GSM#07 Plenary
                           in November 1984   27
       3.1.3   *Memories of the Work on the SMS Concept in the Second Half of 1984*   28
                3.1.3.1   The Overall Expectations and Possibilities in the early
                           80s   29
                3.1.3.2   A Possible Concept for a General Short Message Service   29
                3.1.3.3   Feasibility of a Short Message Service   30
                3.1.3.4   Coding of Characters and Text Formatting   30
                3.1.3.5   Display and Generation of Short Messages   31
                3.1.3.6   Communication Directions   31
       3.1.4   *First Step to a Realisation: The Proposal for Standardisation*
                Submitted to GSM#07 Plenary in Oslo from 25 February to 1 March
                1985   32
                3.1.4.1   The Start of the SMS Standardisation   32
                3.1.4.2   The Proposed Structure and Classification for
                           Standardisation   32
                3.1.4.3   The Finalisation of the Input   33
       3.1.5   *Significance of the Results in the DF900 Cooperation in the Second*
                Half of 1984   33
3.2   The Standardisation of the SMS Concept in the GSM Committee from
       February 1985 to April 1987   34

3.2.1		*Agreement on GSM Service Scope and the SMS Service Concept in the First Half of 1985*	35
	3.2.1.1	Process	35
	3.2.1.2	First SMS Service Concept in CEPT GSM	36
3.2.2		*Elaboration of the SMS Service Requirements*	37
	3.2.2.1	Scope of Work in WP1 from Mid-1985 to Early 1987	37
	3.2.2.2	SMS Work Included in Recommendation GSM 02.03 'Teleservices Supported by a GSM PLMN'	38
	3.2.2.3	A More Elaborate Service Concept and a First Technical Concept	38
	3.2.2.4	Other Service Aspects	39
3.2.3		*Work on SMS Network Aspects from Mid-1985 to April 1987*	41
3.2.4		*Conclusions*	42
3.3		The Acceleration of the GSM Project, Including SMS in 1987	42
3.3.1		*The Decisions of GSM#13 in February 1987 on SMS*	42
3.3.2		*The GSM Memorandum of Understanding (MoU) and its Influence on SMS*	43
3.3.3		*The Beginning of IDEG/WP4 and DGMH*	43

4   The Technical Design of SMS in DGMH from June 1987 to October 1990   45
F. Trosby

4.1		Background	45
4.2		Some Personal Sentiments at the Start	46
4.3		The Instructions that IDEG Were Given for Provision of SMS	47
4.4		Overall Description of the Work in the Period from 1987 to 1990 and Work Items Dealt with	48
4.4.1		*General*	48
4.4.2		*Network Architecture and Transport Mechanisms*	48
4.4.3		*Service Elements*	50
4.4.4		*Supplementary Services*	53
4.4.5		*Short Message Length and Alphabet*	54
4.5		The SMS of September 1990	55
4.5.1		*Some General Remarks on the Way that Services, Networks and Communication Protocols Were Described in the Late 1980s*	55
4.5.2		*The Specifications Defining the Short Message Service Mid-1990*	57
4.5.3		*Network Architecture*	57
4.5.4		*Service and Service Elements*	59
4.5.5		*Protocol Architecture and Service Definition*	60
4.5.6		*Principle Schemes of Short Message Transfer*	61
	4.5.6.1	The Mobile-terminated Case	61
	4.5.6.2	The Mobile-originated Case	61
4.5.7		*Addressing Capabilities*	62

	4.5.8	*Maximum Length of Message*	63
	4.5.9	*Alphabet Available for User Information*	64
4.6		Major Design Issues	64
	4.6.1	*General*	64
	4.6.2	*Inter-MSC Transfer of Short Messages – by X.25 or Signalling Capabilities?*	66
	4.6.3	*Service Centre – Inside or Outside the PLMN?*	67
	4.6.4	*What Interface to Choose for the SMS-SC↔SMS-GMSC/ SMS-IWMSC Connection?*	68
	4.6.5	*Fixed Interworking or Variable Interworking?*	69
	4.6.6	*Some Concluding Remarks*	70
4.7		Final Remarks on the Period of the First Three Years of DGMH	71
4.8		Work on SMS in GSM Bodies Outside GSM4	72
	4.8.1	*GSM 04.11*	72
	4.8.2	*Map Operations for the Support of SMS*	72
4.9		Other Tasks of DGMH	73

5   The Evolution of SMS Features and Specifications from October 1990 to the End of 1996   75
K. Holley

5.1		Topics Discussed in this Chapter	76
5.2		Technical Improvements to SMS 1990–1996	77
	5.2.1	*Continuous Message Flow*	77
	5.2.2	*Multiple Service Centre Scenarios*	77
	5.2.3	*Delivery Reports*	78
	5.2.4	*SMS Character Sets*	79
	5.2.5	*SMS – an Optional Feature?*	79
	5.2.6	*Storing SMS on the SIM Card*	80
	5.2.7	*Unacknowledged SMS*	80
	5.2.8	*Memory Capacity Available*	80
	5.2.9	*SMS Negative Time Zone*	80
	5.2.10	*Length of Binary SMS*	81
	5.2.11	*Sending to and Receiving from Non-Numeric Addresses*	81
	5.2.12	*SMS API*	82
	5.2.13	*Storage of SMS in the Phone*	83
	5.2.14	*Upgrading Network Capability Reduces SMS Length*	83
	5.2.15	*SMS to an External Terminal*	84
	5.2.16	*Specifying Service Centre Interconnect to the Cellular Network*	84
	5.2.17	*Replace Short Message*	85
	5.2.18	*Detecting Terminal Capabilities*	86
	5.2.19	*Nokia Cellular Data Card*	86

	5.2.20	*Improved Error Reporting*	86
	5.2.21	*Receiving SMS from External Systems*	87
	5.2.22	*Avoiding SMS Duplicates*	87
	5.2.23	*Expanding SMS Character Capabilities*	87
	5.2.24	*Using SMS to Alert to Waiting Voicemail*	88
	5.2.25	*Which Time Zone to Use for SMS*	89
	5.2.26	*Delaying SMS after Phone Power-On*	89
	5.2.27	*Manual Flow Control*	89
	5.2.28	*SIM Management by SMS*	90
	5.2.29	*Icons for Voicemail Alert*	90
	5.2.30	*Ensuring Proper Display Control via Control Characters*	90
	5.2.31	*Concatenated SMS*	90
	5.2.32	*SMS Divert*	91
	5.2.33	*SMS Message Indication*	91
	5.2.34	*More on SMS Alphabet Coding*	92
	5.2.35	*Interworking with Email*	94
	5.2.36	*SMS Between Networks*	94
	5.2.37	*Use of AT Commands for SMS*	94
	5.2.38	*SMS Compression*	96
	5.2.39	*More on New Alphabets*	97
	5.2.40	*More on International SMS Messaging*	97
5.3		Concluding Remarks on the SMS Period 1990–1996	97
6		**The Evolution of SMS Features and Specifications from the Beginning of 1997 to Mid-2009**	**99**
		I. Harris	
6.1		SIM Toolkit Data Download and Secure Messaging	100
6.2		SMS Compression	100
6.3		Enhanced Messaging Service (EMS)	101
6.4		Voicemail Management	103
6.5		Routers	104
6.6		Language Tables	105
6.7		Other Important Standards Work for SMS	107
	6.7.1	*Reserved Code Points*	107
	6.7.2	*Port Numbers*	107
	6.7.3	*Support of SMS in GPRS and UMTS*	107
6.8		The End of an Era	108
6.9		Further Reading	109
7		**Early Commercial Applications and Operational Aspects**	**111**
		I. Harris	
7.1		Fixed-network Connection to the SMS-SC	112

7.2   Network Operator Interworking, Roaming and Number Portability   114
7.3   Third-party SMS-SCs   115
7.4   Intelligent Terminal Connections to Mobile Phones   116
7.5   SMS Keyboard Text Entry   117
7.6   SMS to Fax and SMS to Email   117
7.7   Two-way Real-time Messaging Applications   119
7.8   Performance   120
7.9   SMS Traffic Growth   121
7.10   Billing   122
7.11   The Content Provider Access (CPA) Model Deployed in Norway   123
7.12   SMS in 2009   123

8   Global Market Development   125
        F. Hillebrand
8.1   The Creation of a Large Base of Mobiles and the Global SMS Infrastructure   125
8.2   First Use of SMS by Network Operators   126
8.3   How SMS Was Discovered by Young People and Became a Part of the Youth
        Culture and Widely Accepted   126
8.4   SMS Has Become the Leading Mobile Messaging Service and Will Stay in
        the Lead in the Foreseeable Future   127
        8.4.1   Global SMS Traffic and Revenues in 2008   127
                8.4.1.1   Model 1: a Simple Calculation   127
                8.4.1.2   Model 2: Research by a Market Research Firm   128
                8.4.1.3   Model 3: Another Crude Extrapolation   128
                8.4.1.4   Conclusion   128
        8.4.2   Some Considerations about the Market Development   128
        8.4.3   A Forecast for 2013   129
        8.4.4   Long-term Prospects for SMS   129

9   Conclusions   131
        F. Hillebrand
9.1   Factors that Were Critical for the Success of SMS   131
9.2   Proposals for a Further Evolution of SMS: SMS Phase 3   132
9.3   What Can be Learnt from SMS for Standardisation in Other Areas   133
        9.3.1   General Aspects   133
        9.3.2   A Proposal for Optimising MMS by a New, Simple, Ubiquitous MMS   133

Annex 1   Abbreviations Used in Several Parts of the Book   135

Annex 2   Sources for Quoted GSM Documents and Other Documents   139

Annex 3   **Meetings of IDEG/WP4/GSM4 and DGMH in the Period from**
             May 1987 to September 1990   **143**

Annex 4   **DGMH Attendance in the Period from May 1987 to September 1990**   **145**

Annex 5   **Meetings of GSM4/SMG 4 and DGMH in the Period from**
             October 1990 to the End of 1996   **147**

Annex 6   **DGMH Attendance in the Period from October 1990**
             to the End of 1996   **149**

Annex 7   **Evolution of GSM Specification 03.40**   **157**

Annex 8   **Literature**   **165**

Annex 9   **Brief Biographies of the Authors**   **167**

Index   **173**

Introduction

F. Hillebrand

SMS has many applications globally, but personal text messaging, measured in terms of traffic and users, is by far the most common SMS application. We have therefore chosen a title that stresses this personal text phenomenon.

Reasons for this Book

SMS is an incredible global success. At the time of writing (mid-2009) it can be used by all 4 billion GSM customers. More than 50 short messages are generated per month per GSM user. The service has created a $100 billion turnover industry. This success has been enabled mainly by the comprehensive and robust standards that have been made mandatory for every network and for every mobile from the very beginning. The standardisation of SMS started in February 1985 within the framework of GSM standardisation.

We will be celebrating the 25th anniversary of SMS in early 2010. There is an interest in understanding how this service was created, but no comprehensive description of the genesis of SMS is available. Hence, this book will provide a record of the genesis of SMS.

The book is written by people who have been at the forefront of GSM/SMS standardisation work from the time the first ideas of an SMS service were created to the present day. Therefore this book can describe the genesis of SMS from the beginning to present day. A review of the present situation and of market forecasts shows that SMS has a promising long-term future.

Improved Access to the Documents upon which this Book is Based

The authors have attempted as far as possible to base the contents of this book on documentary evidence. The physical and electronic archives of ETSI contain many documents dating back to the early 1990s. However, all temporary documents of the first six GSM meetings, which took place from 1982 to 1984, are missing, and for most subgroups the early documents before the early 1990s are also missing.[1]

In order to fill these gaps, I have retrieved the complete document collection of the early GSM subgroups from the T-Mobile archives. In addition, Mrs Elisabeth Spindler, the widow of the late Dr Klaus Spindler, the head of the German delegation in the first 12 GSM meetings from 1982 to 1986, has handed over to me most of the missing early GSM Plenary documents and documents pertaining to Franco-German cooperation in March 2009. The other missing early GSM documents have been retrieved by Thomas Beijer in the Swedish State Archive in the second quarter of 2009.[2]

This improved documentary base makes it possible to provide a complete and objective description of the genesis of SMS, especially the early phases. Interesting parts of the early documents have been scanned for this book. They are of low quality owing to the limited text processing capabilities available at the time. We have not retyped them, as we wished to retain their authenticity.

Content of this Book

The book starts with an overview of fixed-network communication services and the portfolio of GSM services, in order to place SMS in the context of the developments in telecommunications (Chapter 1). This is followed by a critical review of reports on the invention of SMS in the press and on the Internet (Chapter 2). The development of the SMS concept from mid-1984 to early 1987 is then reported in Chapter 3.

The key technical specification GSM 03.40 'Technical Realisation of the Short Message Service – Point-to-Point' was developed in the DGMH (Drafting Group on Message Handling) in the period from 1987 to 1990 (Chapter 4). The evolution of SMS features and specifications is described in Chapters 5 and 6. The development of SMS features to support applications other than person-to-person communication is described in Chapter 7. Chapter 8 addresses the implementation and market success of SMS, and conclusions are drawn in Chapter 9.

The authors and their Cooperation

Each chapter has a single author whose name is given after the chapter title. All authors have commented on each part. The authors have also sought the comments and views of colleagues.

Friedhelm Hillebrand was strongly involved in the early development of the concept and has written the initial chapters. Finn Trosby, Kevin Holley and Ian Harris were the

---

[1] These documents formed the basis of the book *GSM and UMTS: The Creation of Global Mobile Communication*, which was edited by me.
[2] All these documents are now in the Hillebrand & Partners company archive. The GSM Plenary documents were handed over to ETSI for an update of the DVD 'GSM and SMS archives 1983 to 2000'. It is intended to hand the rest over to a public archive as well.

chairmen of DGMH and the successor group in 3GPP.[3] They provide a complete record of the development of SMS technical design and features and their evolution.

Acknowledgements

We would like to express our sincere thanks for thorough discussions to the former leadership of the GSM committee: Thomas Haug, Philippe Dupuis, Martine Alvernhe, Jan A. Audestad, Bernard Ghillebaert and Thomas Beijer. In particular, Thomas Beijer put a great deal of effort into reviewing the draft manuscripts. Also, Mike Short of O2, former GSM MoU chairman, deserves many thanks for his review and suggestions. Friedhelm Rodermund (former DGMH programme manager) created the footnotes to Chapter 6.

Advice for the Reader

The Terms 'Text' and 'SMS'

'Text communication' is a term that has existed since the 1970s. It broadly describes all forms of text communication (Telex, Teletex, fax, etc.). Mobile network operators in countries like the UK and Finland use 'text' as another expression for 'short message'. In France they use the term 'texto' for 'short message'. It is certainly a better marketing term than 'SMS'. For clarity, in this book we use the term 'text communication' in the broad sense. When we speak about the GSM text messaging service we use 'short message service' (SMS), and when we speak about the message we use 'short message' (SM)'.

As described later on, text messaging functionalities appeared prior and parallel to GSM in many forms (pagers, business platforms such as Mobitex, etc). The GSM service – short message service – comprised elements of these but went much further: a very versatile element for both the private and the business sectors, enabling messaging in both directions and delivery confirmation. The term 'Short Message Service', or 'SMS', was not used by any of the prior systems. GSM succeeded in making it a brand. I think we should maintain it.

How to Find Cited GSM Documents and Other Documents

In this book, GSM documents and several other document types are cited by title and/or number. No documents are reprinted in this book. How such documents can be retrieved is described in Annex 2.

---

[3] See the table in Chapter 3, Section 3.3.3, which provides an overview of the SMS technical standardisation leadership.

1

Communication Networks in the Early 1980s and the Portfolio of GSM Services

F. Hillebrand
Hillebrand & Partners

1.1 Station-to-station Morse Telegraphy, the Origin of All Modern Technical Text Communication

Morse telegraphy, via wires and also wireless, which existed as far back as the nineteenth century, is the origin of all modern technical text communication. It is important to emphasise that text communication over radio is not at all new. It was there when GSM was first defined, it was there many years before and indeed it has always been there for a very long time already. The first radio systems, the Morse systems, were created in the infancy of radio technology, and they were still in use in the 1980s when GSM was defined.

It was actually forbidden to send larger ships across the sea without having equipment and skilled people to handle these systems. Therefore, the importance of radio text communication was fully understood among communications engineers. How to integrate text communication into the new pan-European system was part of the general discussion among land mobile telecom engineers in those days.

These radio services were station-to-station services without any support from any type of 'network'.

1.2 Network-based Communication Services in the Early 1980s

In contrast to station-based communication between two radio stations, network-based communication means that the users communicate with network nodes and pass their information

Short Message Service (SMS)   Edited by Friedhelm Hillebrand
© 2010 John Wiley & Sons, Ltd

via one or several connected nodes to the recipients. This method enables the communication to be supported by, for example, automatic routing, etc.

1.2.1 Telephony

The most widely used fixed-network service was, of course, telephony. The service was fully automated, and practically every household had a telephone. But the fixed networks made progress, and a range of new non-voice services was created. Therefore, it was important to consider these for the conception and standardisation of the new GSM system, which began at the end of 1982.

The first fully automatic mobile communication networks went into operation in the early 1980s. Back then, the number of mobile users was very small compared with the number of fixed users. Mobile communication was seen as an extension of the fixed networks, and hence an objective was that the mobile networks should support the services of the fixed networks, and of telephony above all. Therefore, the scope of the analogue mobile networks existing at that time was primarily telephony.[1] This meant that users could make national and international calls. For calls to mobile users, the callers did not need to know where the mobile was at the time of calling. The mobile network provided an automatic routing of the call. Subscriber numbers were still very small compared with the fixed networks, which was caused by high price levels on equipment, subscription and traffic. People referred to the mobile phone as the 'rich man's toy'. Mobile stations were mounted in cars, with high equipment and service costs. Hence, only professional users could afford them. The situation improved in the mid- and late 1980s. But still the penetration of the population was small.

1.2.2 Telex, the Forefather of Modern Text Communication[2]

The forefather of modern text communication was Telex (= Teleprinter Exchange). It was a public, worldwide text communication service. It allowed dialogue between two 'teleprinters'. This was enabled by the development of suitable terminals with a keyboard and printer function, progress in switching and transmission systems and standardisation of the service. It was first implemented in the 1930s. In 1979 the Telex service was used by 1 100 000 users in 155 countries. It was used heavily for international connections. The 106 000 German Telex users[3] sent 35% of their traffic to foreign destinations in 1976, which accounted for 58% of the revenue.[4] There were extensions of the Telex service via shortwave radio and

---

[1] The first-generation systems were also used for fax and remote control, both entirely based on implementations in the terminals.

[2] Based on information in: Wolfgang Mache (1980), *Lexikon der Text- und Datenkommunikation*. Oldenburg Verlag, Munich, Germany, ISBN 3-486-24361-6, p. 281.

[3] This compares to about 25 000 mobile telephony users.

[4] Norbert Dauth (Ed.) (1978), *Datenvermittlungstechnik (EDS)*. Deckers Verlag, Heidelberg/Hamburg, Germany, ISBN 3-7685-5277-2, pp. 38–39.

satellite services, which allowed text messaging to and from and between ships as well as aircraft (Maritex is an excellent Telex service to ships using the shortwave radio band).

The service was standardised comprehensively, so that it was unnecessary for agreement to be reached between two users before communication (as later in data communication). Instead, the message could simply be sent to any other Telex user by dialling his directory number and thus setting up a connection.

The service had limited possibilities. The bit rate used was just 50 bit/s, which was due to the speed limitations of the partly electromechanical equipment. The text was encoded according to International Alphabet No. 2. This used 5 bits per character and enabled 32 different possibilities to be represented. These were used for 26 letters, without differentiation of lower and upper case, and some control signals (such as carriage return and line feed, space and switching between letters and numbers).

Telex was very successful in spite of the limitation of its service features. It was the backbone of international business communication for many decades. Telex no longer exists as a public service.[5] It is remarkable that it worked globally, was interoperable between equipment of all manufacturers involved and was easy to use. These properties made Telex a great example for the conception and standardisation of OSI (Open System Interconnection) in data communication. Telex was popular with users owing to its simplicity, speed and reliability. Also, Telex messages were considered to be legal documents, very much like letters. It was loved by network operators because it was highly profitable. All this can likewise be said of SMS today.

1.2.3   The Advent of Many Faster Transport Techniques for Text and Data

Telex traffic had been carried on specific 50 bit/s switching and transmission systems. The need to realise higher data rates for more comfortable text communication or for data communication led to new developments:

The *public switched telephone network* was equipped with modems that could offer speeds ranging from 300 to 9600 bit/s in the early 1980s.

New Telex switches and transmission systems with advanced features were developed. This led to a family of *circuit-switched data services* using the CCITT X.21 interface working at speeds of 600, 2400, 4800, 9600 and 48 000 bit/s.[6]

Packet switching services were standardised and developed. The key interface was X.25, offering data rates of 2400, 4800, 9600 and 48 000 bit/s with synchronous transmission. Terminals with asynchronous transmission used a packet assembly/disassembly unit. Possible speeds were 300, 600 and 1200 bit/s.[7]

Progress of technology allowed the building of digital circuit-switched telephone exchanges and digital transmission equipment that was cheaper than the existing analogue

---

[5] There seem still to be company internal applications in aircraft and ship communications.
[6] Friedhelm Hillebrand (1981), *DATEX, Infrastuktur der Daten- und Textkommunikation*. Deckers Verlag, Heidelberg/Hamburg, Germany, ISBN 3-7685-3081-7.
[7] See previous footnote.

equipment. Therefore, the analogue telephone network was converted in steps into a digital telephone network. The analogue speech signals were converted into 64 kbit/s digital signal streams. By adding a digital subscriber line, the ISDN (Integrated Services Digital Network) was created. It offered to users two 64 kbit/s channels for telephony and a 16 kbit/s channel for signalling. This resulted in a total rate of 144 kbit/s. The rate of 64 kbit/s for a voice channel was established long before ISDN in the development of digital transmission systems. The state of the art required such a high bit rate in order to achieve at least the same quality as in good analogue connections.

1.2.4   Creation of New, Fully Standardised Text Communication Services

Facsimile technology was developed early in the twentieth century but did not find a wide use. In 1976 the CCITT approved technical specifications for powerful fax machines (groups 2 and 3 with up to 9.6 kbit/s transmission rates) working in the telephone network. This led to the introduction of public fax services by PTTs in the late 1970s. These fax machines were fully interoperable. Users could simply dial a connection to another fax machine, the telephone number of which could be found in the public directory, and be sure that the message was received and printed correctly.[8] The standardisation of facsimile led to a strong acceptance in the market. Fax was first adopted on a large scale in Japan, as it allowed Japanese characters to be transmitted. This led to a volume production of fax machines. The price of these machines fell, and a breakthrough in the world market was achieved. A large number of communication partners were available. This again drove the usage and penetration of fax machines even to private users.

This success led to the standardisation of group 4 machines intended for ISDN with a 64 kbit/s data rate. However, the machines were expensive, so no high penetration was achieved. Another unattractive element was the interworking with the large base of group 2 and 3 machines. For such connections the speed of the fast machines was reduced and users had no advantage. The use of a server that could store and forward the fax messages and would allow the fast machine to send rapidly was not considered in standardisation. Once again, the lesson was: a wide user base and a low cost make services attractive even if the performance is limited.

Fax was very widely used in the 1980s and early 1990s, but it was superseded by the advent of personal computers and word processing programs such as Word. Received fax messages could not be further processed by a PC. Some people referred to fax messages as 'dead documents'. Therefore, users preferred to exchange text messages as Word documents via data communication or email. Fax is still widely used in many countries. It still fulfils a service need that even email cannot fulfil: the simple 'remote copy' of documents to be sent end to end in real time. A large number of hotels still use fax for reservation purposes.

---

[8] Based on information in: Wolfgang Mache (1980), *Lexikon der Text- und Datenkommunikation*. Oldenburg Verlag, Munich, Germany, ISBN 3-486-24361-6, p. 267

Teletex had been standardised to overcome the limitations of Telex. The initial idea was to convert every office typewriter into a communication terminal. The full character set of typewriters was used. The Teletex messages were formatted as A4 pages, so that a correct representation of letters could be achieved. This was a wonderful idea that needed global standardisation in order to achieve a wide acceptance and volumes of traffic and technical equipment.

A very great effort was made by many network operators and manufacturers in the late 1970s to standardise this new service. As interworking between the different emerging new transport technologies was not possible at the time at reasonable cost, an agreement was needed on a single transport technique. There were fervent advocates of using X.21 circuit-switched digital networks with 2400 bit/s. Then there were other fervent advocates of X.25 packet switching technology and fervent advocates of the telephony network with modems to avoid the cost of a data network. Later the ISDN faction also urged for ISDN offering 64 kbit/s channels for fax transmission. But none of the fighting parties was able to win, and there was no will for a compromise, so they were unable to agree on a single transport technique. Hence, this unresolved dispute led to a segmentation of the market and communication islands that could not communicate between themselves. Several systems became operational in 1981, but they were all switched off during the 1980s.

Then there was an overregulation of the service. A conservative faction requested that, in order to guarantee quality of service, every Teletex terminal be operational for 24 hours 7 days a week. This was possible if a company had one Telex machine in a separate room, but, in instances where all typewriters had been converted into Teletex stations, this requirement was opposed by the people responsible for fire protection. The idea of using a server to store messages to unreachable terminals and to transmit the messages as soon as the terminals became reachable was addressed too late. Such a server was, however, already inherent in the first SMS concept.

In view of all these factors, for which the Teletex designers themselves were responsible, the wonderful service idea of Teletex did not fly in the market. The demand for text communication was later covered by email.

1.2.5 Creation of Value-added Text Communication Services Based on Servers

Interactive Videotex was based on the idea of using a TV set as the user terminal and of offering a dial-in service to a server where information could be accessed. The server stored a lot of information for users and offered access to the remote computers of information suppliers. The users dialled via the telephone network with a low-cost 1200/75 bit/s modem into the server of the Interactive Videotex service. This service found a wide acceptance. However, owing to its restrictions (formats, screen resolution, data speeds, etc.), Interactive Videotex was uncompetitive in the time of wide availability of PCs and was replaced by Internet browsing.

Message Handling Systems (MHSs) were systems where a server supported the generation of the message by a dumb terminal. The MHS stored the message and forwarded it to the target MHS. The recipient could then look into his mailbox and retrieve the message. The MHS could be accessed by a wide range of services: the telephone network with a modem, packet switching networks, etc. Such systems had been standardised since the late 1970s (X.400 series), and the first systems appeared in the market in the early 1980s, providing a service that was similar to email. They were not a great success as a public service, as the cost of access via the telephone network was too high. The generation of a message by a dumb terminal lasted quite some time. Consequently, no operator worked on a low-cost access, e.g. by dial-in ports in every local exchange. Another important reason for X.400 fading into oblivion was definitely the horrendous addressing format of the standard. However MHSs were fairly successful for closed company internal communications. Most computer systems offered such features to their users.

1.2.6   The Development of Private Mobile Radio Networks

According to Wikipedia:[9]

> *Mobitex* is an OSI-based, open, standard, national, public access, wireless, packet-switched data network. Mobitex places great emphasis on safety and reliability with its use by military, police, firefighters and ambulance services. Mobitex was developed at the beginning of the 1980s by the Swedish Televerket Radio.[10]

Mobile text and data networks for private mobile radio applications in the USA and Canada offered in the late 1970s voice and text/data transmission in a single integrated network.

Research work was done well before the standardisation and development phase of such networks and services.

1.2.7   Internet, Web Browsing and Email as the Winners in Communication in Fixed Networks

As a result of the fragmented development in telecommunications, a lack of innovation and a lack of fundamental agreement in telecommunication standardisation in several areas, the Internet has become the present dominant transport network. Fixed text communication has migrated to email, and Interactive Videotex services have been replaced by web brows-ing. The future is looming already in the shape of social networking (Facebook, etc.) and Twitter.

---

[9] Wikipedia, 16 February 2009.
[10] This was confirmed by Thomas Haug, who was involved in Mobitex.

1.3   Services Portfolio of GSM

1.3.1   Way of Working in GSM Standardisation

As this book may be read by people with other professional backgrounds and possibly by students, a short tutorial on how the GSM standardisation machinery worked is offered here. Most people believe that a group of people gather together and start to fight over which of a number of *already developed solutions* to a specific problem is the best. In the end, and after much blood has been spilt, a winner is announced, and that is the standard. People are unaware of the tremendous amount of advanced technical development work that is involved in the standards body itself. This may even be unique to the mobile business, and to GSM in particular.

According to Wikipedia:[11]

> Standardisation is the process of developing and agreeing upon technical standards. A standard is a document that establishes uniform engineering or technical specifications, criteria, methods, processes or practices. Some standards are mandatory while others are voluntary. Voluntary standards are available if one chooses to use them.[12]

For mobile communication systems – intended for a global or pan-European usage – a comprehensive standardisation is needed to achieve the necessary functioning: every SIM card must work in every terminal; every terminal must function in every network; interworking of all networks for message/call routing and for the support of international roaming must be possible.

The standard must enable attractive and economic services and equipment. For this, innovations and inventions are needed. However, for standardisation, not every contribution has to be novel. Often it is sufficient to choose between several well-known solutions and agree on just one, but the end result must be attractive to users, operators and manufacturers.

GSM standardisation work starts with an agreement about strategic targets. A work programme containing several work items is then developed. The work for each work item follows a structured methodology:

- stage 1: service requirements;
- stage 2: architecture and message flows;
- stage 3: interfaces and protocols.

In the early years (up to 1987), decisions required unanimity. They have since been made by consensus. Only in selected cases are votes taken. This open process allows every

---

[11] Wikipedia, 29 March 2009.

[12] ETSI, the European Telecommunications Standards Institute, has very elaborate directives (199 pp.) with many defined terms. However, there is no definition of 'standardisation' or 'standard': 'The objective of the Institute is to produce and perform the maintenance of the technical standards and other deliverables which are required by its members. "STANDARD" shall mean any standard adopted by ETSI' (see ETSI Statutes).

participant to make contributions to all aspects. The process can be described as a joint development process.

To judge the value of work in a standardisation context, it is necessary not just to look at the creation of an idea. For successful standardisation it is necessary to write a contribution paper, to seek support, to submit it to the appropriate standardisation working group, to present it and to make sure that it is accepted and incorporated in an emerging standard's document. In many cases a good solution needs contributions from several parties. Other activities to ensure acceptance are to become an editor of a specification, to monitor the issue over a long period and ensure that it is not removed and to improve it over time. Becoming a chairman and organising the work so that the ideas can flourish is also a possible way of contributing to a successful outcome.

The agreement on a standard is a necessary but not sufficient condition for success in the mobile communication market. There are so many standards that are not used, as they are normally voluntary. Mandatory standards are made obligatory by law. This covers only very restricted areas such as user and network safety. In GSM standardisation, the GSM MoU Group made the GSM standards mandatory for their members. For market success it is important that a high quality standard is actually implemented and widely used. Hence, efforts to seek the agreement of commercial actors to use a standard are very worthwhile if success is to be achieved in the market.

1.3.2   Service Philosophy of GSM Developed from 1982 to 1984

The GSM committee was founded at the end of 1982, and it was given the task of standardising a future pan-European mobile communication service and system using the 900 MHz band. The GSM committee agreed upon the following as basic targets for the service:

- The focus was to be upon mobile telephony, but it was expected that non-voice services would also be required.
- Mobile stations were to be used in all participating countries, including support of handheld mobile stations.
- All services that existed in the fixed networks were to be made available to mobile stations.

The existing analogue mobile telephony networks had nearly no non-voice services. Neither could they offer SMS-like services. Packet radio networks had supported text communication since 1978. Also, private mobile radio systems (e.g. Mobitex) had supported text messaging since about 1980. Therefore, a need existed to offer attractive non-voice services in the GSM system, the next generation of mobile networks.

The 'coexistence between vehicle-borne and handheld mobile stations' led to a long discussion. No existing European mobile system supported handhelds. In Central Europe there was very limited spectrum available for public mobile services, as much of the spectrum was used during the cold war for military purposes in many countries. Therefore, delegates feared that insufficient capacity would remain for stations onboard cars, trucks, railways and ships where a radio connection was indispensable. However, all delegates saw that the support of handheld stations would be attractive to customers. Therefore, it was concluded that the GSM standard should allow handhelds to be introduced at the discretion of the network operator. This meant that the GSM standard should allow small cells to be implemented and should specify special functions to enable cheap handhelds.[13]

1.3.3 GSM's Fixed-network-service Companions

1.3.3.1 Reference Model

A reference configuration was agreed for all services that were defined as companions of the fixed-network services (Figure 1.1).

The reference configuration envisaged the same specification for the terminals (TE in the diagram) in the fixed and mobile networks. Other unstandardised terminals could be adapted by using a terminal adapter function (TA). Within the mobile network, specific functions are needed to ensure a sufficient quality of service. This includes a special channel coding and, in some service, specialised error correction protocols. These functions could not be understood by the fixed networks. Therefore, an interworking unit (IWU) was needed between the mobile and fixed network. The special functionalities would be handled by the TE or TA on the one side and the IWU on the other. This made it possible, in an elegant way, to concentrate the additional functions needed for non-voice services in defined units and to

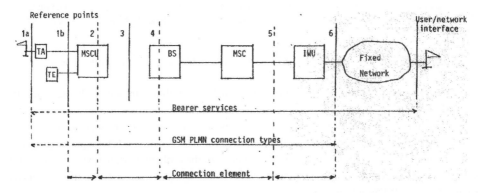

Figure 1.1 Reference model for mobile communication[14] (TA = terminal adaptor; TE = terminal equipment; MSCU = mobile station central unit; BS = base station; MSC = mobile-services switching centre; IWU = interworking unit

---

[13] GSM provides the following functions which help handhelds: power control and discontinuous transmission and reception to save battery consumption, and slow frequency hopping to enhance the speech quality in a fading environment.

[14] GSM document 28/85 rev. 2, Information for Industry, p. 8.

install these according to demand without creating additional cost for the basic telephony service.

1.3.3.2 Teleservices

Teleservices were defined at the time as: 'A type of telecommunication service that provides the complete capability, including terminal equipment functions, for communication between users according to protocols established by agreements between administrations'.[15]

A wide range of teleservices was envisaged:

- short message transmission;
- access to message handling service;
- interactive videotex;
- Teletex;
- Telex;
- fax;
- teleaction service (alarm service, telemetry service, etc.);
- picture transmission.

The Short Message Service, as a very special type of messaging implemented as an integral part of the signalling systems, was proposed in GSM as the only new service that did not already exist in public networks (Figure 1.2).

> Advantage has been taken of the radio aspects of the system in utilizing it to provide a short message point to point or multipoint Teleservice without call set up, which is not available on fixed networks.

Figure 1.2   First description of SMS as a new teleservice in a standard document[16]

Efforts were made to interest the people responsible for ISDN in standardising it jointly so that a seamless service would exist. But there was no interest on the fixed side.[17] The reason for this was probably that the ISDN representatives saw mobile services as a small brother proposing a service with severe restrictions for customers (message length) that they did not want to impose on their customers, as they planned rich text services such as Teletex. But they did not understand that there was a mass market for a simple service usable on every ISDN telephone.

In the family of envisaged SMS services, only the point-to-point services were really commercially successful. The point-to-multipoint services, later called cell broadcast, were

---

[15] GSM document 28/85 rev. 2, Information for Industry, p. 9.

[16] GSM document 28/85 rev. 2, Information for Industry, p. 3.

[17] I had talks in the late 1980s at a time when we were seeking a harmonisation of supplementary services. I undertook a second imitative when I was ETSI TC SMG chairman in the debate about fixed-mobile convergence in ETSI in the late 1990s.

not successful, as no attractive applications were found and no reasonable business model could be developed.

Hence, the SMS point-to-point services remained the only new services that had no companion in the fixed networks. It became a very successful GSM service. This book will explain how and why.

Substantial work was put into the standardisation of an access to message handling systems, especially by the Nordic operators. A full technical report was developed in standardisation. The service required a dumb terminal, all intelligence was in the MHS server and hence lengthy dialogues took place on the circuit-switched GSM bearer services used for access to the MHS. This was too expensive, and consequently it was not accepted by the market.

The standardisation of a mobile Teletex version was completed, but it was not implemented, as the service died in the fixed networks. Mobile Telex was seen as too old and was therefore not standardised.

The mobile fax version was standardised, but it was not successful in the market because it required clumsy fax machines connected to mobile stations and because the fax service was superseded in fixed networks during the first half of the 1990s when the GSM systems were launched.

The teleaction and picture transmission services were discussed at the time in the fixed network standardisation. However, the standards were not completed, and the services were not implemented in fixed networks and hence not further considered in GSM.

1.3.3.3 Bearer Services

Bearer services were defined at the time as: 'A type of telecommunication service that provides the capability for the transmission of signals between user–network interfaces'.[18]

A wide range of bearer services was envisaged:

- Telephony.
- 3.1 kHz audio band signals.
- Circuit-switched data services:
  - unrestricted digital (i.e. raw channel of about 13 kbit/s);
  - 300 bit/s duplex asynchronous;
  - 1200 bit/s duplex asynchronous;
  - 1200/75 bit/s duplex asynchronous;
  - 2400 bit/s duplex synchronous;
  - 4800 bit/s duplex synchronous;
  - 9600 bit/s duplex synchronous;
  - several half-duplex versions.

---

[18] GSM document 28/85 rev. 2, Information for Industry, p. 9.

Mobile telephony became the most successful service of GSM. The 3.1 kHz audio band signals bearer service did not find much market interest. All the circuit-switched data services (except half-duplex) were standardised and implemented by most networks. Many had high hopes for CS-based mobile data, and were very enthusiastic when 14.4 kbit/s and High-Speed CSD came around. However, this range was not very successful in the market, for many reasons:

- The call set-up time was way too long.
- Even high-speed data were not fast enough for Internet browsing.
- Reliability was a problem. There were dropped calls all the time, caused by the poor radio coverage early on.
- In the early years, laptops were pretty clumsy and heavy.
- Moreover, the charges per session were pretty high because most IT applications required a dialogue between terminal and host computer, where transactions took place only from time to time. The channel was therefore not used most of the time, but the charging counters counted the entire connection time. The network operators could have applied some sort of volume charging, as the function of DTX (Discontinuous Transmission) did make sure that radio resources were only used when data were sent.

Around the year 2000, the GPRS (General Packet Radio Service) in GSM took these applications over.

1.4 GSM Mobile Telephony and SMS – the Most Successful Telecommunication Services

GSM/UMTS reached 3552 million users worldwide at the end of 2008[19] (Figure 1.3).

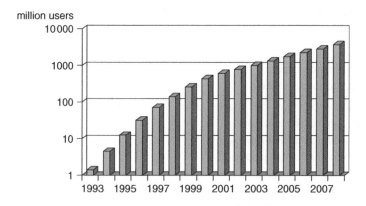

Figure 1.3   Number of GSM and UMTS users worldwide[20]

---

[19] GSMA wireless intelligence. *Quarterly World Review*, Q4 2008.

[20] Data from www.gsm.org, own graphic.

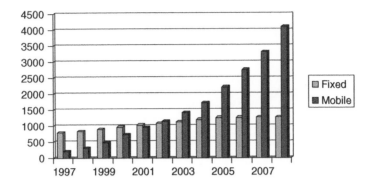

Figure 1.4   Main telephone lines and mobile users worldwide, in millions[22]

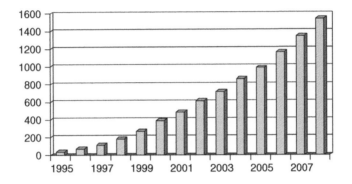

Figure 1.5   Number of Internet users worldwide, in millions[23]

GSM/UMTS was preferred by 89.5% of all mobile users at the end of 2008.[21] Therefore, it is the dominant standard for mobile communication.

By 2002, mobile telephony had already overtaken fixed telephony in the number of subscribers worldwide (Figure 1.4).

The number of GSM subscribers is greater than the number of Internet users (Figure 1.5): 3552 million versus 1542 million in 2004. Hence, more people can use SMS than email.

It is estimated that the world market saw about 3–4 trillion short messages and a revenue of the order of $80–100 billion in 2008. A substantial growth in the number of messages can be expected in the coming years. The growth in revenue will be slower. More details can be found in Chapter 8, Section 8.4.

---

[21] Source www.gsacom.com

[22] Data from: http://www.itu.int/ITU-D/ict/statistics/at_glance/KeyTelecom99.html, own graphic.

[23] Data from: http://www.itu.int/ITU-D/ict/statistics/at_glance/KeyTelecom99.html, own graphic.

Figure 1.4 Mobile telephone network and GSM subscriber numbers, in millions.

Figure 1.5 Forecast: the global market for mobile terminals.

GSM/UMTS was replaced by GERAN as all work in the current 3GPP. Therefore, it is the dominant standard for mobile communication.

By 2002, mobile telephones had already overtaken fixed telephony in the number of subscribers worldwide (Figure 1.4).

The increase of GSM subscribers is greater than the number of subscribers for fixed GSM million across 1992 and was no small. The one wrote photos set out 94% their usable that was noted but the world market saw at an 1.4 million new discovered and awesome of the value of approximately billion in 2005. A substantial growth in the number of terminals can be expected in the coming years. The growth in revenue will be slower. More multimedia can lead to reduce in revenue slot.

Data from figures and ITU Publications: 2004 and ITU 1994, ITU 1999, ITU 2002, ITU 2004.
Data comparison is an end. UMTS replaces the 3G market by growth in the in 2005, their evolution.

2

Who Invented SMS?

F. Hillebrand
Hillebrand & Partners

2.1 Introduction

There have been many reports in the press about the 'invention' or 'discovery' of SMS. The terms 'invention' and 'discovery' are used often in a very loose way. In order to understand what has happened, what was claimed and which claims are justified, it is necessary to talk first about these terms and gain a precise understanding before dealing with the matter further in this chapter.

2.2 Clarification of the Terms 'Invention' and 'Innovation'

2.2.1 Invention

An invention must be novel and not obvious to those who are skilled in the same field, i.e. not only to persons who have no specific knowledge in that field:

> An invention is the creation of a new configuration, composition of matter, device or process. Some inventions are based on pre-existing models or ideas. Other inventions are radical breakthroughs that may extend the boundaries of human knowledge or experience.[1]
>
> An invention shall be considered to be new if it does not form part of the state of the art. The state of the art shall be held to comprise everything made available to the public by means of a written or oral description, by use, or in any other way ....[2]

---

[1] Wikipedia, 16 February 2009.
[2] http://www.epo.org/patents/law/legal-texts/html/epc/1973/e/ma1.html, paragraph 54.

Short Message Service (SMS)   Edited by Friedhelm Hillebrand
© 2010 John Wiley & Sons, Ltd

2.2.2   Innovation

According to McKeown:[3]

> The term innovation means a new way of doing something. It may refer to incremental, radical and revolutionary changes in thinking, products, processes or organisations. A distinction is typically made between invention, an idea made manifest, and innovation, ideas applied successfully.

According to Wikipedia:[4]

> Those who are directly responsible for application of the innovation are often called pioneers in their field, whether they are individuals or organisations.

2.3   Was SMS Invented during the ISDN Work?

The thesis that SMS was invented during the ISDN work was asserted for quite some time by people involved in the standardisation of network aspects of the ISDN and GSM.

ISDN is a circuit-switched multiservice network. In later phases of its standardisation (in the early 1980s), a packet-switched communication between users began to be created using the signalling channel of the ISDN access. At the time of the start of GSM SMS standardisation, this feature was not very developed and was coupled to the circuit-switched connection in order to provide a second communication channel. The solution was also relatively complicated, as this was a packet-session-oriented concept without any limitations on the message length. The signalling channel (D-channel) allowed data to be transmitted at a rate of 16 kbit/s. This is a much higher bit rate than the rate of the signalling channels in GSM. Therefore, a full OSI protocol could be used, with no need for message length limitation. This packet switching service over the ISDN service does not envisage the use of a 'message handling services centre' that could treat cases of non-reachable users/terminals.

During the phase of the SMS service specification, we did not know how to transport short messages in the fixed network. We approached the people responsible for ISDN standardisation in several countries and tried to find a solution for a short message service that would span over GSM and ISDN. But they were not interested, and nothing happened.

Therefore, there is no service comparable with SMS standardised in ISDN, and the term 'short message (service)' is not used in this arena at all.[5]

---

[3] Max Mckeown (2008), *The Truth About Innovation*. Pearson/Financial Times, London, UK.

[4] Wikipedia, 29 March 2009.

[5] There is, however, a type of SMS service available in ISDN now that is based on a non-standardised solution in selected cordless phones and specific servers. It has limited functions. It uses a modem in the telephony channel for SMS data transmission. Every first ringing tone of the phone is suppressed. The phone listens whether a data connection is requested. If it hears the modem tones, it will not ring, but transmit the short message.

2.4 Was SMS Invented by Test Engineers, Students or in a Pizzeria Session?

The overwhelming market success of SMS led the press after 2000 to show great interest in finding out about the origin of SMS. As newspapers want strong stories with heroes, they searched for 'the inventor of SMS'. It seemed to them to be unthinkable to have such an important development and no inventor who might become rich by royalties. Two different stories became known: the SMS discovery/invention by test engineers or students and the Copenhagen pizzeria story.

2.4.1 Was SMS Discovered or Invented by Test Engineers or Students?

This storyline developed in the UK. On 5 December 2002, the tenth anniversary of SMS was celebrated by the 160 Characters Association. Articles appeared about the first transmission of a short message in December 1992.[6] In December 2007, the 15th anniversary of SMS was celebrated and the story of the first SMS was repeated.[7]

Some newspapers went further and described the event of the first message transmission as an occasion where a hidden possibility of the GSM system was discovered or even invented.[8]

A Finnish newspaper claimed, however, that the first short message was sent in Finland.[9] The Finnish journalist also mentioned that there were rumours that Nokia engineers had discovered the possibility of sending messages. In addition, she reported about a book publication: 'The technology had been invented by Finnish schoolboys who were nervous about asking girls out for dates ... and Finnish schoolgirls who wanted to tell each other about what happened during the dates ...'.[10]

The sending of the 'first short message' cannot be interpreted as the invention of SMS because SMS was not created by the act of sending the first message. Instead, SMS required much effort to develop concept ideas, standardise the service and implement it in many networks.

2.4.2 Was SMS Invented in a Copenhagen Pizzeria Session in 1982?

There was a series of reports in European newspapers about the invention of SMS. The source of these reports was a publication in *Helsingin Sanomat* in a special monthly supplement called *Kuukausiliite* in May 2002. The journalist had found a Finnish engineer who

---

[6] http://instantmessagingplanet.com/wireless/article.php/1553321
[7] http://www.160characters.org/news.php?action = view&nid = 2471
[8] For example, http://morgenpost.de/printarchive/magazin/article157852/SMS
[9] *Helsingin Sanomat*, May 2002; see also Section 2.6.
[10] Michael Lewis (2002), *The Future Just Happened*. WW Norton, New York/London, ISBN-10 0393323528.

had reported that he had discussed with colleagues in a Pizzeria in Copenhagen in 1982 the possibility of transmitting texts between mobile stations. This general idea had not been further elaborated by methods, techniques, configurations, etc. They had neither documented their idea nor provided contributions to standardisation.

The journalist interpreted this as the invention of SMS. This story found its way into Wikipedia and even led to an innovation prize from *The Economist* in October 2008.

However, this idea was not novel, as a broad range of text messaging services had been standardised and in operation before the date of the Pizzeria session:

- There had been Morse telegraphy radio transmissions of text since before 1900.
- In the 1970s, Telex, a worldwide text messaging service, provided text messaging to, from and between mobile stations onboard ships or aircraft.
- Packet radio technology for text and data communication was first used by radio amateurs in Canada in 1978.
- Mobile text and data networks were first used in Sweden (Mobitex), the USA and Canada in the late 1970s–early 1980s.

The content of the invention claimed by the journalist was just the exchange of text between mobile stations, and the idea had not been specified further by methods, techniques or configurations. This idea was known in the state of the art and hence was not novel. Therefore, it was incorrect to call it an invention.

2.5 A Clarifying Discussion within the GSM Community in Spring 2009

As a reaction to the various press reports, a discussion within the former leadership of the Technical Committee GSM took place in order to set the record straight. There were also discussions with the participants of the Pizzeria session mentioned in Section 2.4. The participants Thomas Beijer, Philippe Dupuis, Thomas Haug, Friedhelm Hillebrand, Kevin Holley, Matti Makkonen, Seppo Tiainen and Finn Trosby agreed as follows:

> The increased international interest in knowing the roots of the present highly successful mobile services has often focused on one specific service, SMS, the Short Message Service. The success of SMS has been a surprise to many, and its global popularity is a natural reason for the interest in its origin.
>
> We, ex-colleagues from the time when cornerstones of the present modern mobile communications were created, have studied together the state of the art in telecommunications and old documentation from the European Standardisation Committee GSM preserved in the ETSI archive. Based on the study and our discussion, we agree on the following:
>
> 1. Text messaging was a known telecommunications service years before the development of GSM, the Global System for Mobile communications, started in 1982 as pan-European cooperation.

2. Proposals for text messaging as a service in GSM were made by the cooperating Nordic operators as well as by the cooperating German and French operators.

3. The Nordic operators focused their work on text messaging by using an access to a message handling system, a service similar to email. This service was standardised by the GSM committee. It led to a technical report on the technical realisation of access to message handling systems.

4. The German and French operators focused their work on 'Short Message Transmission'. This proposed service uses a dedicated services centre and transmits the text messages over existing signalling paths of the GSM telephony system on a lower-priority basis. This transmission method obviously restricts the message to be short – maximum length initially estimated as 128 octets, later optimised to 160 characters, still sufficiently long for most personal or professional purposes. The concept was developed in the Franco-German cooperation programme in the 1983/4 timeframe.

5. In a first work phase from February 1985 to the end of 1986, the GSM committee specified the service features of the Short Message Service (SMS). Most contributions came from Germany and France.

6. From 1987 onwards, the technical realisation of SMS was standardised in a small group called Drafting Group on Message Handling. The first chairman of this group and the editor of the key technical specification were provided by Norway (later replaced by the UK), and technical work was mostly provided by Finland, France, Norway and the UK. The first phase of the SMS specifications comprised items such as service definition, network architecture, topology and protocols, acknowledgement capabilities, functionality for alerting on messages waiting, time stamping and capabilities of identifying application protocols.

7. The further evolution of the SMS was standardised in the same small group led by the UK (where the technical work was mostly provided by the UK). Examples of enhancements from this period are automatic replacement of messages, so-called 'flash SMS' and voicemail icons, followed by colour and picture capabilities and long SMS.

Our conclusion is that the origin of the text messaging services in GSM can be found in the historical development of telecommunication services. SMS was created by a small group of people. Work on the standardisation of services and technical realisation was approved by the GSM committee dealing with the standardisation of the new Pan-European Mobile Communication Service and System.[11]

This document was reviewed by ETSI and the substance was published by ETSI on their website.[12]

2.6   Timetables of SMS Genesis

2.6.1   Concept Development and Standardisation of SMS

---

[11] Document sent on 20 April 2009 to ETSI by F. Hillebrand.

[12] http://www.etsi.org/WebSite/Technologies/Cellularhistory.aspx

Table 2.1 Concept development and standardisation of SMS

Period	Area	Work	Body	Achieved milestone
Mid-1984 to Feb. 1985	Development of service concept	Development of first SMS service concept	Franco-German cooperation	SMS service concept
Feb. 1985 to end of 1986	Standardisation of the person-to-person service definition	Development of the SMS service definitions	GSM WP1	GSM 02.03 Teleservices of Jan. 87 in WP1 70/86 rev.2 or in IDEG 07/87
Mid-1987 to end of 1990	Technical standardisation of SMS person-to-person. Phase 1	Technical design of SMS phase 1 (see Chapter 4)	GSM IDEG DGMH	GSM 03.40 version 3.4.1
		Support of SMS on Mobile Radio Interface (see Section 4.8.1)	GSM WP3	GSM 04.11
		Support of SMS in the network (see Section 4.8.2)	GSM WP3, SPS SIG	GSM 09.02 MAP
End of 1990 to 1992	Enhancements to SMS person-to-person. Phase 1	Technical design of SMS phase 1	GSM WP4 DGMH	GSM 03.40
		Support of SMS on Mobile Radio Interface	GSM WP3	GSM 04.11
		Support of SMS in the network	GSM WP3, SPS SIG	GSM 09.02 MAP
1992 to 1996	Technical standardisation of SMS person-to-person. Phase 2	SMS phase 2	GSM WP4 DGMH	
		Enhancements of the support of SMS on Mobile Radio Interface	GSM WP3	GSM 04.11
		Enhancements of the support of SMS in the network	GSM WP3, SPS SIG	GSM 09.02 MAP
1996 to 2005	Technical standardisation	Selected new features		

2.6.2   Development of SMS in the Market

Table 2.2   Development of SMS in the market

Milestone	Achievements
1992	First acceptance tests by various operators in Europe
1993	First SMS point-to-point mobile-terminated-enabled phones available
	First uses by network operators for alerting of received voice mails
1994	Every new terminal was capable of SMS point-to-point mobile-terminated
1995	Every network was capable of SMS
	International roaming for SMS available
1995	Discovered by youngsters, and began to become a part of the youth culture
1996	Every new terminal was also capable of SMS point-to-point mobile-originated
	National interworking between operators was in place between most operators
	Substantial traffic
2008	3–4 trillion short messages sent, with a revenue of $80–100 billion worldwide

3

The Creation of the SMS Concept from Mid-1984 to Early 1987

F. Hillebrand
Hillebrand & Partners

3.1 The Birth of the SMS Concept in the French and German Network Operators

3.1.1 Introduction

As shown in Section 2.5, the roots of SMS creation were in the work of the operators of France and Germany. In order to meet the huge short- and medium-term demand for mobile telephony, a cooperation programme started in mid-1983 with the target of developing and implementing a common 'interim' analogue 900 MHz mobile communication system (S900 cooperation programme) that was to cover the demand until GSM became available. A start of operation was envisaged in both countries in 1986.

In mid-1984, both parties made the decision not to pursue the idea of an analogue non-standardised interim system but to go directly for the standardised pan-European digital mobile communication system GSM. They agreed 'to introduce before the end of the present decade a common digital cellular radio communication system based on a European standard, the definition of which is presently under way in CEPT'.[1] They planned as a first step a joint experimental programme for digital mobile communication techniques. The new target required much more intensive joint work to prepare contributions to the GSM standardisation work. This second cooperation programme with the GSM focus was called the DF900-cooperation programme.

---

[1] 'Objectives' section of the Franco-German agreement of 30 October 1984, see GSM document 76/84.

Short Message Service (SMS)   Edited by Friedhelm Hillebrand
© 2010 John Wiley & Sons, Ltd

The joint work was steered by the DF Project Group. It was jointly chaired by Philippe Dupuis (France) and Dr Klaus Spindler (Germany). The minutes of these meetings and most meeting documents have survived.[2] This group worked mainly on strategic matters, of course, but there are some relevant traces of SMS-related aspects (see Section 3.1.2). Then there were small subgroups dealing with service aspects, radio aspects and network aspects. The subgroup dealing with services was jointly led by Bernard Ghillebaert (France) and Friedhelm Hillebrand (Germany). The documents of this group did not survive.[3]

In Section 3.1.2, all relevant documentary evidence that has survived will be considered. In Section 3.1.3, memories of the programme leaders are reported. Section 3.1.4 describes the preparation of the input to GSM Plenary in February 1985, where the standardisation of SMS started. Section 3.1.5 discusses the significance of the results.

3.1.2   Documentary Evidence that Survived

After so many years, it is difficult to find documentary evidence. The search is sometimes like archaeology. Often only small traces have survived, which need to be put in context and interpreted. The totality of the pieces found results in a fairly good mosaic of the development.

3.1.2.1   Roots of SMS in 'Enhanced' Paging Integrated in a Multiservice Mobile Communication System

After standardisation of the fixed ISDN (Integrated Services Digital Network) in the late 1970s there were also ideas in many minds to integrate several mobile services into a new public digital mobile communication network.[4] Advanced PMRs (Private Mobile Radio networks) and PAMRs (Public Access Mobile Radio networks) integrated several services such as speech, text and data in one network (e.g. the PAMR network Mobitex, which was conceived in about 1981). The idea of a single network supporting several services was from the beginning part of the GSM concept at the end of 1982: 'It is expected that, in addition to normal telephony traffic, other types of service (non-speech) will be required in the system'.[5] Such statements were based on inputs from the Netherlands and Nordic countries in the meetings preparing the way for the first GSM meeting.

A radio paging service was proposed as an integrated feature of the future pan-European mobile communication system in CEPT working group SF (Services and Facilities) in March 1983 (Figure 3.1).[6]

---

[2] The documents should be available in the archives of Orange France and T-Mobile Germany. Copies are available in the company archive of Hillebrand & Partners.

[3] Participants were Christian Bénard-Dendé and Ulrich Grabolle, but the principal work on SMS was done by Bernard Ghillebaert and me. Unfortunately, again, the personal notes of all participants have not survived.

[4] Such concepts were discussed early in the Nordic cooperation. It was mainly driven by Jan A. Audestad.

[5] See GSM 02/82, Section 3c.

[6] Quoted from 'Summary of definitions' taken from the SF handbook in document GSM 14/83, March 1983, put together by Thomas Beijer, Sweden, and submitted to GSM in March 1983, Section 2.4. This definition is also found in GSM 38/83 of October 1983.

Radio Paging Service

This service gives to the user the possibility of receiving personally from any telephone in a public network by radio, notice, with or without a message, wherever he may be in a given area.

Figure 3.1   Proposal for a radio paging service in the future GSM system

GSM asked SF for a clarification and definition,[7] but no further input was received on this matter from SF.

The French 'Marathon' project, a research project for a new digital mobile communication system, proposed to integrate alongside the telephony service a data service, a store-and-forward mailbox service and an evolved paging service. This was described in a technical paper in November 1982.[8] The evolved paging service is characterised as 'a service of a fast distribution of short messages on radio paths (type Eurosignal[9] 'evolved', with or without confirmation of the reception (depending on the technical and economical feasibility))'. The 'evolution' dealt with the greatest problem of paging services. Existing paging services sent out the message by radio, often repeating it once, but there was no confirmation from the recipient to the sender that the message had been received.

3.1.2.2   A Surprising Proposal in the S900 Interim System Context

A proposal can be found in a French input document to the S900 programme of 12 September 1983.[10] In Section 7.6 of this document, a service is described (Figure 3.2).

7.6 Short data messages

> This service could be offered to a special category of mobile stations equipped for data transmissions. Short data messages should be transmitted on special signalling channels and message queuing should be used for messages to mobile stations.

Figure 3.2   Proposal for a short data messages service in the Franco-German interim system S900

---

[7] GSM 53/83, Section 2.6.

[8] Projet Marathon, Note Technique, November 1982. This document should be available in the archives of Orange France. A copy is available in the archive of Hillebrand & Partners.

[9] Eurosignal was a Europe-wide paging service.

[10] Functional Specification for a Cellular Radiotelephone System Common to Several Countries, document S900–16, Section 7.6. This document is a draft version of the functional specification. It is probably the first draft and may therefore be considered as a French proposal. The mention of DTM/DS on the front page means that Philippe Dupuis was the editor of this document.

It might seem astonishing that such a feature was proposed for an analogue mobile communication system, but modern analogue mobile systems like the C450 network in operation in Germany and the S900 design used a core network that was based on a digital switching system and had digital control channels separated from the speech transmission channels throughout the system. These systems would have allowed the implementation of a short data message service.

The service proposed in the section cited above is unidirectional to the mobile user and is limited to mobile stations that have been specifically equipped. Hence, this proposal is a variant of the enhanced paging services integrated into a mobile communication system.

However, such a complicated new feature was not compatible with the necessary time plans for an interim system that was to cover the mobile telephony demand until GSM became available. Nobody thought of SMS as a function for the S900 interim system. Hence, the idea was deleted in the final version of the request for proposals.[11]

Nevertheless, the text shows that many elements of the later SMS proposal were state-of the art thinking at an early time.

3.1.2.3   The Mutation of an Enhanced Paging Service to a General Short Messaging Service in the Specification Process of the DF900 Trial Systems for GSM

The joint experimental programme planned by France and Germany for digital mobile communication techniques envisaged several projects: 'Each project concerns the design, manufacturing and supply of experimental models of radio subsystem elements of a 900 MHz digital mobile communication system. By experimental model is meant equipment making it possible to conduct experiments in laboratories and field sites. These experiments will aim to compare different technical solutions applicable to the radio subsystem'.[12]

For the technical specification of the trial system in the coordination meeting on 29–30 October 1984, France proposed the inclusion of an enhanced paging service (Figure 3.3).[13]

```
- provision of an enhanced "paging" service, i.e.  diffu-
  sion of alphanumeric messages to mobile users with
  acknowledgement  capabilities  (i.e.  assuming  duplex
  transmission).
```

Figure 3.3   Requirement for an enhanced paging service in a GSM trial system in October 1984

The document was discussed in the meeting and changed at a request from Germany from an 'enhanced' paging service to a general (short) message transmission service (Figure 3.4).[14]

---

[11] Information from Philippe Dupuis who has retained the French version of the request for proposals.
[12] Agreement between France and Germany of autumn 1984.
[13] DF900-3 General Technical Specifications (for the trial systems), submitted by France to the DF900 meeting on 29–30 October 1984.
[14] Handwritten correction by Dr Klaus Spindler in his copy of the document, clean typed version in a facsimile from FTZ to CNET on 8 November 1984; a copy is available in the H&P company archive.

                        a message transmission
   – provision of ~~an enhanced "paging"~~ service, i.e. diffu-
 sion of alphanumeric messages to mobile users with
 acknowledgement capabilities (i.e. assuming duplex
 transmission).

Figure 3.4   Change of the requirement for an enhanced paging service into a two-way messaging system in a GSM trial system in October 1984

The technical specifications for the trial systems were discussed in the DF900 coordination meeting on 12 November 1984 in London, again in order to finalise them for the request for proposals to be sent to industry in due course. There, again, the draft input document of France was modified at a request from Germany, consistently with the former changes (Figure 3.5).[15]

      – the transmission of short alphanumeric messages ~~to mobi-~~
    ~~le users~~ with acknowlegement capabilities (i.e. assuming
    duplex transmission),

Figure 3.5   Change of the requirement for an enhanced paging service into a two-way messaging system in a GSM trial system in November 1984

Paging is by definition of the ITU a one-way service. Therefore, the term had to be changed in order to express the two-way functionality properly. By these changes, the integrated enhanced paging service, which was directed from the fixed network to a mobile subscriber, was changed to a general short messaging service that could be mobile-terminated and mobile-originated. It could support communications between mobile and fixed users as well as between mobile users.

This modified text was included in the request for proposals for the trial systems that was sent to industry in early December 1984. This shows the agreement between France and Germany not to include an enhanced paging service but to go for a general short message service.[16]

3.1.2.4   A Technical Discussion and Input to GSM#07 Plenary in November 1984

There was an intensive discussion about the channels on which packet data should be carried. Frieder Pernice[17] proposed not to use associated control channels for this purpose, as they would have a capacity of 100–300 bit/s only. If these channels were upgraded to the needs of packet data transmission of several kbit/s, the spectrum efficiency of the system would

---

[15] Handwritten correction by Dr Klaus Spindler in his copy of the document, now in the H&P company archive.
[16] Contained in the request for proposals sent to industry by FTZ on 7 December 1984, Part 2, Technical conditions, Section 2.2.3; a copy is available in the H&P archive
[17] Head of department in the German Bundespost's Technical Engineering Centre.

be impaired.[18] This ended the discussion on using associated control channels for general-purpose packet data transmission. The subsequent discussion in the DF900 programme concentrated on the use of associated control channels for short messages only.

A reflection of the discussions and internal studies can be found in an input document of France to the GSM#06 meeting in November 1984.[19] It discusses principles for layer 2 of the interface between mobile station and base station and states that 'MS_BS link protocols may be established for signalling or for some kind of data application such as short message transmission'. This contribution was briefly discussed by Working Party 3 Network Aspects and then integrated into the meeting report for further study and agreement.[20] Here, the term 'short message transmission' is used but not explained. This means it must have been familiar to the authors of the contribution. The Working Party 3 meeting obviously did not discuss the input document thoroughly, or else a definition would have been added.

Two documents from the DRM conference in Espoo in early February 1985 (before the GSM Oslo meeting) mention enhanced paging by short message transmission. Their authors are from Germany and France.

3.1.3   Memories of the Work on the SMS Concept in the Second Half of 1984

Both in France and Germany there had been study work on new digital mobile communication systems since the early 1980s. From the end of 1982 onwards, both countries participated in GSM work. Some discussions on GSM work had already taken place during the S900 cooperation project.[21] With the agreement on the new focus for Franco-German cooperation on the GSM system in mid-1984, joint preparatory work for the GSM committee was strongly intensified.

In Germany we had a number of preparatory internal meetings.[22] The Franco-German DF900 Project Group had an intensive discussion about GSM standardisation in a meeting in September 1984.[23] Further discussions took place during the Franco-German Project Group meetings in 1984 and 1985.[24]

But separate working meetings on service aspects also took place.[25] Some dates can be found for these meetings, but no minutes or meeting documents have survived.[26] Most

---

[18] Letter of F. Pernice to P. Dupuis and B. Ghillebaert of 7 November 1984. A copy is available in the archive of Hillebrand & Partners.

[19] GSM document 75/84, Section 4.

[20] GSM document 90/84, Appendix 1 to Annex 5, p. 33.

[21] The Project Group meeting of 13–14 September 1983 mentions in its report a discussion of GSM-related matters (S900–21). The Project Group meeting of 10 October 1984 considered the GSM programme and DF cooperation for the GSM work (see meeting report).

[22] In my agenda, I found meeting dates in Darmstadt on 29–30 August, 26–27 September and 19 October 1984.

[23] S900 meeting #10 on 18 September 1984 in Bonn.

[24] The highest-level group managed by Philippe Dupuis and Dr Klaus Spindler.

[25] Managed by Bernard Ghillebaert and Friedhelm Hillebrand.

[26] 12 November 1984, London, at the occasion of the discussion of specifications for trial systems; 12 February 1985 in Paris.

team members had a background in text and data communications. As no documents have survived, I would like to report from memory on the work in the services working group.[27]

3.1.3.1 The Overall Expectations and Possibilities in the early 80s

The best-known study on demand for mobile services was elaborated by PACTEL in 1981 for the EURODATA Foundation.[28] It forecast 5–20 million mobile users for Europe in the year 2000.[29] This means about 1–3 million users in the larger European countries. This was only a fraction of the number of fixed telephony users. And with the high number of trucks, buses and executive cars, this means that only a small fraction of the number were hand portables. This picture was consistent with the possibilities. As the cold war was still going on, most of the envisaged 900 MHz spectrum was in use by the armies of many countries that were assembled in Europe.

3.1.3.2 A Possible Concept for a General Short Message Service

The strategic targets of the GSM committee made it clear that the GSM system needed to be optimised for telephony as its most important usage. The cost for telephony needed to be low in order to reach the mass market. There was an interest to have attractive non-voice services to make the new system future proof. However, the size and kind of demand was unclear. Therefore, no advance investments for non-voice users could be justified. In this context, our thinking on possible non-voice services began. We discussed a range of services. The following text concentrates on what became the SMS.

A key success factor for telecommunication services is a large number of possible communication partners.[30] Hence, it was envisaged to equip every mobile station for SMS, to enable market success.

To do this practically without additional cost, features existing in the mobile station for telephony purposes, such as the keyboard and display, would need to be used. For message transmission, the signalling links needed for telephony could be used. Such links had a limited throughput. Owing to the limited quality of the underlying radio links, the protocols used frames with a length that would allow only the transmission of a short message. Also, SS#7, the signalling system used within the fixed part of the mobile network and between mobile and fixed networks, had a maximum frame length of less than 300 bytes, of which only a part was usable for short message transmission. For a simple solution, no

---

[27] The result as reported here has been reviewed in a brainstorming session between Bernard Ghillebaert and me in London on 24 March 2009.

[28] The EURODATA Foundation had been created by the European data network operators in order to publish detail about data communication services available in the European countries. It also published data user numbers and traffic that existed and forecast such numbers.

[29] Results quoted in DF900 trial systems request for proposals 11 November 1984. First mentioned in the DF meeting on 30 October 1984, in document DF900-3.

[30] This was experienced in the 1970s when several new services were introduced. The most striking example is fax. Fax traffic took off only after the rigid standardisation of group 3 machines and their wide availability.

session concept with a breaking down of long messages into a sequence of packets could be envisaged. This led to the technical possibility of designing a short message service that would be able to send and receive short messages with acknowledgements. The maximum message length was estimated in the beginning as 128 bytes, allowing the transmission of about 146 characters using 7 bits per character of International Alphabet No. 5.[31]

The only new network element needed was a dedicated server to store and forward the short messages. This server could also take care of the delayed transmission of messages to mobile stations that were temporarily unreachable. Initially there was the hope that this could be done by a standardised message handling system. However, during the standardisation it turned out that a dedicated short message service centre was the optimal solution (see Section 3.2.2.3).

3.1.3.3   Feasibility of a Short Message Service

The key question, however, was whether such a restricted service offering only the possibility of exchanging such short messages would be useful to future mobile users at all. No possibility of market research existed. Intensive discussions led to two convincing positive plausibility arguments that it was possible to create a useful service in spite of the severe limitation on message length:

- In the consumer market, postcards with a photo or drawing on one side and an address and text field on the other side were very popular. Many postcards carried less than 146 characters.
- In the professional market, Telex was at the time the most widely used text messaging service. More than half of all Telex messages were shorter than 146 characters. Telex used a circuit-switched connection and allowed a dialogue between the two machines while the connection was established. Therefore, a lot of short messages existed in the Telex service. Also, many fax messages were short at the time.

3.1.3.4   Coding of Characters and Text Formatting

As code for the characters, the International Alphabet No. 5 using 7 bits per character was a natural choice, as it was widely used in new telecommunication services.[32] For short messages, page formatting would be useless. Hence, a simple line-oriented transmission was seen as sufficient.

---

[31] This size could be enhanced to 140 bytes carrying 160 characters later in the standardisation work (see Section 4). In the mid-1990s it became possible to create longer messages and transport them by concatenated short messages. See Section 5.

[32] The parity bit (eighth bit) was not needed, and neither were the control characters – hence the 7-bit default alphabet for SMS. Control character positions were populated with European national variants.

3.1.3.5   Display and Generation of Short Messages

The short messages could be shown on the display which the receiving mobile station needed anyhow for telephony functions. Such displays were not very large at the time, but the text could be scrolled to be read.

The generation of a short message by a mobile station was not so easy. The use of an external, fully featured keyboard connected to the mobile station was clumsy and hence unattractive.[33] Telephone keypads at the time, in several countries, had letters besides the numbers on the keys, as is the case today in mobile stations. These were used to dial 'names' of subscribers. Such keypads were standardised. They could be used to type the characters of short messages by repeated typing of the same key. This was not comfortable. Our French colleagues were sceptical as to whether this was practical for wide usage in the market. I thought that for a mobile user with no other option it might be acceptable. We all underestimated the capability of young users to play virtuoso on a standard telephone keypad, as well as the creativity of developers (T9 system[34]).

Messages coming from the fixed network could be created by a terminal with a fully featured keyboard connected to the fixed network. In France, the Minitel terminal and service was used in practically every household. This service was introduced initially to replace the telephone directory book by an online retrieval of telephone numbers. Many other applications were added to this system over time. The creation of a short message to be sent to a mobile station could be easily added. In most other countries an interactive videotex service standardised by CEPT existed that could also be easily expanded for the generation of short messages to be sent to mobile stations. The interactive videotex services have now been replaced by the Internet, but it is still possible to send short messages from an Internet terminal to a GSM mobile station.

3.1.3.6   Communication Directions

The long existing idea of an enhanced paging service integrated into the new mobile communication system could be realised by the short message service 'point-to-point, mobile-terminated'. A message could be generated with a Minitel terminal or Interactive Videotex terminal and be sent to the SMS server and from there to the mobile station. This service is actually a sort of enhanced paging with a technical confirmation for the receipt of the message. Also, the message length was longer than in most existing alphanumeric paging systems. The thinking in France started from this service, as can be seen in Section 3.1.2.

In Germany we were thinking of a more general text messaging service, where mobile users could also generate and send a short message. This message could be put on the server used by the SMS for retrieval by fixed users. The server could transmit the short message to another mobile user.

---

[33] Later, for several mobiles, attachable keyboards were developed (e.g. by Ericsson).

[34] See Section 6.

3.1.4   First Step to a Realisation: The Proposal for Standardisation Submitted to GSM#07 Plenary in Oslo from 25 February to 1 March 1985

3.1.4.1   The Start of the SMS Standardisation

The idea of a general short message service was included as a requirement in the technical specifications of the DF900 trial systems, as described in Section 3.1.2.3. In order to ensure the realisation of SMS in the GSM system, it was decided by the French and German operators to propose SMS to GSM for standardisation.

The GSM#06 Plenary in November 1984 had announced that the GSM#07 meeting from 25 February to 1 March 1985 in Oslo would focus on services and facilities. There had already been WP1 meetings on services during the plenaries in Rome and Gothenburg. These were ad hoc meetings overseen by the Chairman of the CEPT group SF (Services and Facilities). But progress has been slow. The original expectation that the services work would be done by CEPT group SF did not come about, as SF was overloaded with fixed network subjects and the participants had a fixed network background and no specific knowledge about mobile services.[35] Therefore, GSM planned Working Party 1 Services during this Plenary to process the expected input. Martine Alvernhe (France) was appointed Chairman during the GSM Plenary in Oslo. WP1 Services met regularly from then on and was later confirmed by GSM as a permanent working party. There had been WP1 meetings before the Oslo meeting, but they had the mandate to study the viability of hand portables. Therefore, the WP1 meeting in Oslo was the first meeting of the WP1 services group.

3.1.4.2   The Proposed Structure and Classification for Standardisation

To be successful, telecommunication services need robust functioning. The possibility of sending messages to a very large number of users requires a complete and robust standardisation. We thought it would be best to follow the methodology developed in ISDN, namely first to define the short message transmission as a teleservice (stage 1) and then to develop the architecture and message flows (stage 2) and finally the detailed protocol specs (stage 3).

After some discussion, we came to a joint proposal with three different services for short message transmission (Table 3.1).

Table 3.1   Proposal for three services for short message transmission

Configuration	Origination/termination	Classification
Point-to-point	Mobile-terminated	Essential
	Mobile-originated	Additional
Point-to-multipoint	Mobile-terminated	Further study

---

[35] See GSM 14/83, 34/83, 38/83, 53/83, 25/84, 37/84.

In order to ensure broad implementation, we used a classification method that was developed, to my recollection, in the ISDN context. A service can be classified as 'essential', which means that it is mandatory for every network and mobile station. If it is classified as 'additional', then network operators can choose whether they implement it in their networks. However, if it is implemented, the standardised specification needs to be applied.

We agreed between France and Germany that the SMS 'point-to-point, mobile-terminated' should be classified as 'essential'.

However, we could not agree between France and Germany on a classification as 'essential' for mobile-originated possibilities owing to differing views on the suitability of the clumsy input mechanism for a mass market. Therefore, this service was classified as 'additional', which meant that it would be fully specified in the standard. If a manufacturer or operator chose to use this variant, they would need to be compliant with the specifications.

3.1.4.3   The Finalisation of the Input

We had a meeting to finalise our input on 12 February 1985 in Paris.[36] In order not to flood the delegates, we decided to create a relatively short description of the SMS within the comprehensive input document GSM 19/85 which covered all proposed services and principles for them.

This document introduced the idea of a short message service for the first time to the GSM committee. In order not to create resistance, we described just the short message transmission point-to-point mobile-terminated and brought the rest of the material to the WP1 meeting that took place during the GSM meeting.

This document was the first of a long series of GSM contributions prepared jointly by Bernard Ghillebaert and me. Philippe Dupuis used to call them the 'Ghillebrand' documents.

3.1.5   Significance of the Results in the DF900 Cooperation in the Second Half of 1984

Probably none of the elements of the common SMS proposal to GSM was novel. This is certainly true for the inputting of text into a mobile station, the transmission of a text message to another mobile station and the appearance of the message on the display of the recipient. The concept of an integrated enhanced paging service was not novel, as it was mentioned in CEPT groups in March 1983.[37] Neither was the data transport on signalling links novel, as this existed, for example, in the specifications of ISDN user–user signalling. Hence, none of the elements of the SMS proposal can be called an invention. However, the combination of the elements in the SMS package was probably novel.

---

[36] DF900#2 report, Section 6.
[37] GSM 14/83 and 53/83.

The SMS proposal developed in the DF900 cooperation in the second half of 1984 had the objective of creating a service that would be useful to customers. Therefore, not only was a service concept proposed, but also a process for standardisation and implementation was started. The first steps were:

- to build a consensus of two large European network operators[38] to implement SMS;[39]
- to include the concept as a requirement in the technical specification of the DF900 trial systems;
- to agree on a proposal for standardisation of SMS as a teleservice in the GSM system; and
- to propose a classification as essential/additional, which would ensure a very wide application.

This type of work/achievement can be called a proposal for an innovation, as defined in Section 2.2. But the work of many more people and many more contributions were necessary to make it work and hence to enable the innovation.

3.2 The Standardisation of the SMS Concept in the GSM Committee from February 1985 to April 1987

The GSM committee (chairman Thomas Haug since the end of 1982) had reached an agreement on strategic questions such as the service philosophy, targets for service quality, spectrum efficiency and the support of hand-portable devices in the period from the end of 1982 to the end of 1984.

In November 1984 the GSM committee had agreed to establish an experimental programme for the new digital radio transmission techniques and the low-bit-rate speech coding, the two most critical new technologies needed for the GSM system. Members were invited to offer experimental systems. These systems needed to be developed and tested. Comparative evaluations of the candidates were planned for the end of 1986, and a decision for the beginning of 1987. After this decision, a phase of detailed technical design and elaboration of technical specifications was foreseen.

The period from early 1985 to early 1987 was necessary for the fundamental radio and speech coding work that would assure a future-proof system. In parallel, the period was utilised to elaborate and agree service requirements and principles of network aspects. In the GSM#06 meeting held in London in November 1984 it was agreed to dedicate a good part of the time of the next meeting in February 1985 to services and facilities.

---

[38] France and Germany represented a market potential of more than 130 million inhabitants, equivalent to about 40% of the total potential in the countries working together on the GSM system.

[39] This carries even more weight in view of the fact that France and Germany, as the only operators in Europe at the time, had agreed in the cooperation agreement to implement a GSM system.

3.2.1 Agreement on GSM Service Scope and the SMS Service Concept in the First Half of 1985

3.2.1.1 Process

In GSM#07 held from 25 February to 1 March 1985 in Oslo,[40] contributions from the Nordic operators on the telephony service and related supplementary services were submitted.[41] For text messaging, a service based on an access to MHS was proposed by them.[42]

The operators of Germany and France proposed a structure of teleservices and bearer services aligned with ISDN definitions and a broad portfolio of services including a new service called 'short message transmission'. The configuration of sending a short message from the fixed network to a mobile subscriber was described.[43]

The GSM committee formed a subgroup responsible for the definition of services and facilities (Working Party 1). The chairperson was Martine Alvernhe (France). This group met during the plenary week. Germany and France explained the full SMS concept in this meeting.[44] Working Party 1 was able to reach an agreement on a first definition of the services for the GSM system. In the text messaging area, this included short message services and access to message handling systems. This GSM service scope, including the first SMS service concept,[45] was approved preliminarily as the basis of further work by the GSM committee in the Oslo meeting.[46]

The next plenary meeting, GSM#08 from 10 to 14 June 1985 in Paris,[47] had again an embedded WP1 meeting. It revised the tables of the report of WP1 approved by GSM in the meeting in February 1985 in order to bring the terminology into line with existing definitions. Also, the table on supplementary services was revised.[48]

As GSM deemed it necessary to inform the manufacturers, a version of this document dated 'GSM Paris June 1985' and marked 'Information for Industry' was derived from

---

[40] Meeting report in GSM 44/85.

[41] GSM 07/85 and 08/85.

[42] GSM 08/85, Section 3.8.

[43] GSM 19/85, Section 3.3.

[44] In the personal notes of Dr Klaus Spindler it can be seen that the mobile-originated SMS was also mentioned in the plenary meeting. These notes are now in the Hillebrand & Partners company archive.

[45] The Plenary report says that the WP1 meeting report is in Doc. GSM 28/85 and that there were two revisions. This document 28/85 rev. 2 (date February 1985, location Oslo) was retrieved by Thomas Beijer from the Swedish State's Archive. It will be published on the updated ETSI DVD at the end of 2009. The document 28/85 rev. 2 contained in the existing version of the ETSI DVD is the version GSM doc. 28/85 rev. 2 (date June 1985, location Paris) distributed to industry after GSM#08 in June 1985.

[46] Meeting report in GSM 44/85, p. 9.

[47] Meeting report in GSM 120/85.

[48] GSM meeting report 120/85, p. 6; the result document is in a version with the number 28/85 rev. 2 (date June 1985, location Paris). This version is available in the Hillebrand & Partners company archive. It is proposed to be included in an update of the ETSI DVD.

the WP1 paper and distributed to industry for information about the envisaged service scope.[49]

3.2.1.2   First SMS Service Concept in CEPT GSM

The 'Information for Industry' contains information on SMS (Figure 3.6).

> **Advantage has been taken of the radio aspects of the system in utilizing it to provide a short message point to point or multipoint Teleservice without call set up, which is not available on fixed networks.**

Figure 3.6   General characterisation of the short message service[50]

This text shows that the GSM committee realised that a completely new service not available in the fixed network was to be created in the GSM system.

The table on teleservices contains more detailed information (Figure 3.7).

Explanations were given at the end of the table (Figure 3.8).

The SMS is completely integrated into the GSM services portfolio. This means that many features defined for the whole portfolio are applicable to SMS as well (for examples, see Section 3.2.2).

Tele-service N°	Type of user information	Information transfer and access attributes	Higher layer attributes	general attributes		comments E/A
				Interworking possibilities	Minimum of quality of service (90% of area and time)	
1 short message trans-mission 1.1. 1.1.1. 1.1.2. 1.2.	short text messages (e.g. 128 octets)	point to point MT MO point to multipoint;MT	FS FS	MHS (Note 1)	FS e.g probability of correct message reception	No call set up E A FS
2.						

Figure 3.7   Detailed information about short message transmission[51]

---

[49] The cover letter is contained in GSM 66/85 retrieved from the Swedish State's Archive by Thomas Beijer. It will be included in the update of the ETSI DVD. The attached document GSM 28/85 rev. 2 'Services sand facilities to be provided in the GSM system', marked 'Information for Industry', dated June 1985, location Paris, is found on the ETSI DVD under GSM 28/85 in the folder of GSM#07 (the previous meeting in February 1985). It will be correctly stored on the updated ETSI DVD.

[50] Page 3.

[51] Page 14.

```
E : Essential ie must be provided in each PLMN
A : additionnal ie provisions will be made for
    these services in the GSM standards ; operators
    may choose to implement them as required
FS : Further study
MO : Mobile Originated calls
MT : Mobile Terminated Calls
QOS : Quality of Service
```

```
Note 1 : Message handling systems in
fixed networks, including paging sys-
tems.

Note 2 : If 3.1 kHz audioband signals
connection type provided.
```

Figure 3.8   Explanations[52]

This condensed first agreed description contains the following important elements: point-to-point SMS mobile-originated and mobile-terminated, point-to-multipoint mobile-terminated. It envisages the use of a message handling system in the fixed network (possibly including the fixed part of the GSM network). The solutions for use as a standardised MHS turned out not to be feasible during the technical design work (see Section 3.2.2.3). The switches in the GSM core network were standard ISDN switches. Therefore, no store-and-forward functionality existed within any of the network nodes MSC, BSC and BTS. As messaging required a specific store-and-forward functionality, it turned out that a dedicated SMSC (Short Messages Service Centre) became necessary.

In addition, the classifications of 'essential' or 'additional' for the point-to-point services are given. The definitions of 'essential' and 'additional' are also included.

3.2.2   Elaboration of the SMS Service Requirements

3.2.2.1   Scope of Work in WP1 from Mid-1985 to Early 1987

During the period from mid-1985 to early 1987, GSM Working Party 1 elaborated service descriptions for all GSM services, including SMS, and created the recommendation[53] GSM 02.03 'Teleservices' in the GSM systems.[54] I was editor of this specification during this period. WP1 created GSM recommendations on bearer services and supplementary services. In addition, WP1 also addressed a broad range of other service-related topics:

- types of mobile station and features of mobile stations;
- quality of service and security requirements;
- subscription and charging;
- international accounting for international roaming;
- free circulation, type approval and licensing of mobile stations.

The specific work on SMS concentrated on service requirements. It was embedded in the work for all services and therefore did benefit from this work.

---

[52] Page 15.

[53] Technical specifications were named 'recommendations' at that time, as they were not binding to the participants in standardisation.

[54] IDEG document 07/87, retrievable from ETSI server closed bodies: go to SMG, then GSM4.

3.2.2.2   SMS Work Included in Recommendation GSM 02.03 'Teleservices Supported by a GSM PLMN'

The first version of recommendation GSM 02.03 'Teleservices Supported by a GSM Public Land Mobile Network (PLMN)' was established in September 1985.[55] The service descriptions for all GSM services were completed and inserted into GSM 02.03. It contained a datasheet per service. The contributions to the SMS work in WP1 came from France and Germany.

For SMS the use of International Alphabet No. 5 was planned.[56] The information transfer was described as packet connectionless. The description of the services concentrated on the communication between the mobile station and the message handling system. It was required that the received messages be confirmed.

It was specified that a mobile-originated short message would need to go to the message handling system. Thus, the message handling system became the hub of the short message communication. How mobile-originated short messages could be further routed to a fixed or mobile user was not described.

During the WP1 meeting in The Hague from 2 to 4 April 1986, recommendation GSM 02.03 was reviewed and updated by approximate values for several parameters, contained in [ ] to show that they were preliminary:

- Maximum message length: [100] characters.
- Probability of correct message reception: better than [99.5]%.
- Maximum transmission delay: [10] s.[57]

These figures set a framework for the technical design; in particular, a relatively high maximum delay value was proposed in order to allow a simple technical design of SMS.

The term 'short message point-to-multipoint mobile-terminated' was replaced with the clearer term 'short message cell broadcast'.

3.2.2.3   A More Elaborate Service Concept and a First Technical Concept

In November 1986, France and Germany submitted a document 'Definition of the short message service' concerning the mobile-terminated point-to-point service to GSM Working Party 1 Services Aspects and to Working Party 3 Network Aspects.[58] The service-related items were processed by WP1 in their meeting in November 1989.

WP1 considered this input in the Berlin meeting on 4–7 November 1986. A small editing group extracted the service-relevant aspects and proposed an annex to the datasheet of the short message teleservice point-to-point mobile-terminated in recommendation

---

[55] GSM WP1 12/85.

[56] As standardised by ITU CCITT Recommendation T.50.

[57] See meeting report in WP1 21/86, Annex 5, based on input from D and F in WP1 16/86.

[58] WP1 66/86, WP3 96/86.

GSM 02.03.[59] It was sent to WP3 for comments and included in the recommendation.[60] It contained a number of requirements:

- Confirmation of the use of International Alphabet No. 5.
- The requirement that users need not know the location of the mobile station they wish to send a message to. The routing should be done automatically by the network.
- Two levels of message acknowledgement: reception of the message by the destination mobile station and retrieval of the message by the user.
- A short message service centre that needed to be connected to the land-based part of the GSM network was assumed instead of a general-purpose message handling system to control message delivery.
- The user should be able to choose between two delivery options: immediate delivery (e.g. within 5 minutes) or delivery within a given time period (e.g. 24 hours).
- Needs for international operation were identified:
  - a user in country A wants to send a message to a mobile user of country A roaming in country B;
  - a user in country A wants to send a message to a mobile user of country B.

The concept was well received. The major conclusion was an agreement that it was not practical to modify a standardised MHS centre, but that a dedicated SMS centre was needed, as the SMS requirements were too different from the MHS requirements.

Some photographs of the GSM WP1 meeting in Berlin in November 1986 are shown in Figures 3.9 to 3.11.

Bernard Ghillebaert (F) participated in the meeting, as shown in the list of participants, but he does not appear in the photos. He may have been outside the room when the photos were taken.

3.2.2.4   Other Service Aspects

The subscription in GSM 02.13[61] was defined without differentiation in respect of the services included. This means implicitly that all services classified as E (essential in every GSM network) and A (additional = fully standardised application according to operator's choice) would be accessible by the 'basic' subscription. This avoids a barrier to the use of SMS by the need for a special subscription.

The WP1 work on charging and international accounting could be applied in principle to SMS.[62]

The detailed work on 01.06 'Service Implementation Phases'[63] was fully applicable to SMS.

---

[59] WP1 86/86, Annex 5.
[60] WP1 70/86 rev. 2 'Status of WP1 Recommendations' of 26 January 1987.
[61] See GMS 02.13 in WP1 46/86, 59/86, 70/86 rev. 2 (of January 87) and 36/87.
[62] Started with WP1 53/86 rev. 1.
[63] WP1 15/87, 37/87, Section 6.9 and Annex 10.

Figure 3.9 GSM WP1, Berlin, November 1986 (from right to left): Martine Alvernhe (Chairman, F), Christian Bénard Dendé (F), unknown, Ullrich Grabolle (D), Friedhelm Hillebrand (D), Hans Tiger (Permanent Nucleus) and Batu Karasar (SF)

Figure 3.10 GSM WP1, Berlin, November 1986 (from left to right): Martine Alvernhe (Chairman, F), Boudewijin Huenges Wajer (NL), Marcel Meijer (NL), Helene Sandberg (N), Bengt Sköld (S), Ian Germer (UK) and David Barnes (UK)

Figure 3.11 GSM WP1, Berlin, November 1986 (from left to right): Friedhelm Hillebrand (D), Ulrich Grabolle (D) and an unknown delegate

3.2.3   Work on SMS Network Aspects from Mid-1985 to April 1987

GSM Working Party 3 was responsible for network aspects of all services. They worked closely together with CEPT SPS/SIG, a group specialising in ISDN network protocols. They were heavily loaded to define the new network for GSM telephony. They dealt with critical items such as mobility management, security procedures, handover procedures and numbering/addressing/identification. In a systematic way they developed the principles for:

- The network architecture: identification of network elements, reference configurations and network functions, and connection types.
- Specification of the interface between the mobile station and the base station (general aspects, layer 2 and layer 3).
- Specification of the interface between the base station system and the switching system (general aspects, layer 2 and layer 3).
- Specification of the mobile application part, the protocol between mobile service switching centres and between networks.

It is therefore no surprise that not much time was left for issues that were not part of the telephony service, such as SMS or other data services. However, WP3 envisaged the transmission of user information in addition to signalling information in the protocols of the specified interfaces.

The mobility management developed by WP3 for telephony was a good basis for the automatic routing of short messages to roaming users.[64]

The technical aspects of the proposal mentioned in Section 3.2.2.3 could not be treated in WP3 Network Aspects, as this group was overloaded with the specification of GSM telephony-related network aspects. Therefore they were considered in 1987 by the newly created IDEG/WP4 (see Chapter 4).

3.2.4   Conclusions

The standardisation phase from February 1985 to April 1987 resulted in a general concept of SMS. Selected aspects of the service were described. The progress was similar to that in other non-voice service areas. Owing to the broad range of subjects addressed, there was a lack of contributions and resources in WP1 and WP3. Hence, priority was given to telephony aspects. SMS therefore became a 'Sleeping Beauty' in this period.

Thorough technical work with more resources was necessary for progress to be made. The necessary technical SMS work was soon enabled by decisions concerning the basic parameters of the standard and the creation of the GSM MoU, which provided a new drive and offered additional resources.

3.3   The Acceleration of the GSM Project, Including SMS in 1987

3.3.1   The Decisions of GSM#13 in February 1987 on SMS

During the GSM#13 meeting in February 1987, the basic parameters of the GSM standard were agreed. This decision confirmed the optimisation of the system for telephony and finally decided that the GSM system should be a digital system. The key parameters of the digital radio transmission system were agreed. A low-bit-rate speech codec was selected, which was optimised for speech only (and not for modem signals and DTMF transmission).[65]

This decision made it necessary to seek a solution for circuit-switched data using terminal adapters and interworking functions:[66] 'As a consequence of this decision, GSM agreed on the necessity of putting extra resources on the task of defining solutions for the definition of terminal adaptors, etc.'[67] In view of the fact that WP3 Network Aspects had informed GSM that they were overloaded and could not deal with the work for non-voice services in the required timescale, France, Germany and the United Kingdom proposed a new Working Party 4 for non-voice services.

GSM agreed as a first step to establish a new 'sub-working party on data services' in principle, and Friedhelm Hillebrand (Germany) was appointed chairman. He was asked to agree terms of reference with the WP3 chairman Jan A. Audestad. They agreed to

---

[64] As requested in WP1 66/86.
[65] GSM 82/87, Section 6 c, second paragraph.
[66] GSM 39/87.
[67] GSM 82/87, Section 6 c, third paragraph.

call the group IDEG (Implementation of Data-services Experts Group) and to make it a separate working group reporting to GSM and working in close cooperation with WP3. This agreement was endorsed by GSM#15 on 12–16 October 1987 in London.[68] The IDEG was later called Working Party 4.

3.3.2   The GSM Memorandum of Understanding (MoU) and its Influence on SMS

The Memorandum of Understanding on the Implementation of a Pan-European 900 MHz Digital Cellular Mobile Telecommunication Service by 1991 was signed on 7 September 1987 by 14 network operators from 13 European countries.[69]

The memorandum of understanding (MoU) contained the commitment to procure the necessary equipment and to open service in 1991 based on the GSM standard. This gave the voluntary standard elaborated in the CEPT GSM committee a binding character for all signatories of the MoU. The signatories also committed themselves to provide the necessary resources to complete the standards in time.

The signatories organised a GSM MoU Plenary and working groups and an action plan. The GSM MoU organisation took over from the GSM committee the recommendation GSM 01.06 'Implementation Phases', IPR Matters and Charging and Accounting Matters. This work was refined. Also, a more detailed implementation plan for the networks and services was elaborated. This all provided direction, energy and resources to the work of the GSM committee and hence was also beneficial for the SMS work as well.

3.3.3   The Beginning of IDEG/WP4 and DGMH

IDEG met for the first time from 20 to 22 May 1987 in Bonn. In the second meeting in June 1987 in Heckfield, UK, a subgroup DGMH (Drafting Group on Message Handling) was installed to deal with the access to MHS and SMS. The first chairman was Finn Trosby of Telenor. With this group, finally an appropriate working structure for technical SMS work existed that could awake the 'Sleeping Beauty' SMS.

An overview of the leadership for the technical SMS standardisation work is given in Table 3.2.

In March 2005 the work of 3GPP TSG-T was amalgamated with 3GPP TSG CN to produce 3GPP TSG CT. The work of T2 was initially allocated to CT2, but, as the MMS development work had been transferred to OMA and there was no ongoing SMS work, it was decided subsequently to merge CT2 into the newly created CT1 (which had responsibility for core network signalling). The work of CT1 was dominated by the work of the previous CN1 group, and only two or three ex-T2 delegates attended CT1. In order to ensure that the

---

[68] GSM 200/87, Section 7d, p. 9.

[69] The text can be found in the CD ROM (folder D) attached to the book *GSM and UMTS: The Creation of Global Mobile Communications* (edited by F. Hillebrand).

Table 3.2 Leadership of the groups dealing with the standardisation of SMS and editorship of GSM 03.40

Role		From	To	Name
Chairman of 'mother' data working group	CEPT/ETSI: • IDEG • WP4 • GSM4 • SMG4	March 1987 April 1988 October 91 June 1992 February 1997[1]	April 1988 October 1991 June 1992 February 1997 June 2000[2]	Friedhelm Hillebrand Graham Crisp Michel Krumpe Wolfgang Roth Kevin Holley
	3GPP: TSG2 WG2	January 1999 May 2001	May 2001 February 2005	Kevin Holley Ian Harris
Chairman of technical group dealing with SMS	CEPT/ETSI GSM: DGMH	June 1987 September 1990 May 1997 March 1999	September 1990 February 1997 March 1999 June 2000[3]	Finn Trosby Kevin Holley Arthur Gidlow Ian Harris
	3GPP T2 SWG3	March 1999 September 1999 May 2001	September 1999 May 2001 February 2005	Arthur Gidlow Ian Harris Josef Laumen for MMS and Ian Harris for SMS
Editor GSM 03.40		June 1987 September 1990 February 1997	September 1990 February 1997 Ongoing	Finn Trosby Kevin Holley Ian Harris

[1] Kevin Holley was confirmed as Chairman by Technical Committee SMG in February 1997. He chaired already the December 96 the closing Plenary, since Wolfgang Roth had to leave early.
[2] The SMG4 group was terminated during the SMG#32 meeting in June 2000.
[3] The DGMH group was terminated during the SMG#32 meeting in June 2000.

expertise of T2 was not lost, an informal group of specialists from the old T2 group was established that to this day has maintained a core of expertise dating back to the origins of SMS in the late 1980s.

In GSM#15 in October 1987 the first action plan of IDEG was presented to GSM. It contained the new recommendation GSM 03.40 'Technical Realisation of the Short Message Service', which is the core technical specification for SMS.[70]

---

[70] Annex 7 provides an overview of the genesis and evolution of GSM Specification 03.40.

4

The Technical Design of SMS in DGMH from June 1987 to October 1990

F. Trosby
Telenor Mobil

4.1 Background

By spring 1987, the technical work in the GSM bodies had reached the stage where it was time to go ahead with the detailed specifications of the non-voice services. After some discussions on how best to provide for rapid progress in this area, it was decided to establish a new group, IDEG (Implementation of Data and Telematics Expert Group) under the chairmanship of Friedhelm Hillebrand from Detecon.

The first meeting of IDEG was held in Bonn in May 1987. The meetings of this group during the first three years are listed in Annex 3 'Meetings of IDEG/WP4/GSM4 and DGMH in the Period from May 1987 to September 1990'. The attendance of the DGMH meeting in this period is given in Annex 4 'DGMH Attendance in the Period from May 1987 to September 1990'.

To propose a draft mandate and its course with the design of non-voice services, IDEG based its work on a set of documents from the GSM main body and from WP1.

The chairman of IDEG proposed to establish a number of subgroups – or so-called drafting groups – to concentrate on different subsets of the services and functionalities for which IDEG was responsible. The normal working procedure of a drafting group was to be assembled at a period of time allocated for parallel sessions. The outcome of the drafting sessions would then be dealt with by IDEG in plenary at the end of the meeting.

At the second meeting of IDEG – Meeting 2 in Heckfield, UK, July 1987 – it was decided to establish a drafting group for the different messaging services that were listed for GSM.

Short Message Service (SMS) Edited by Friedhelm Hillebrand
© 2010 John Wiley & Sons, Ltd

The categories of objectives were twofold:[1] (i) designing a set of services for the sending and receiving of short messages; (ii) designing a set of services for the access to the MHS (Message Handling System). At the end of the meeting in Heckfield, I was appointed chairman of the DGMH (Drafting Group on Message Handling), which was to start its activity at the next IDEG meeting.

Abbreviations used are explained in Annex 1. Sources for quoted GSM documents can be found in Annex 2. Many documents have been used to prepare this chapter. They can be downloaded from www.gsm-history.org.

4.2   Some Personal Sentiments at the Start

The work in IDEG/WP4/GSM4 represented my first participation in work with GSM, and the assignments as chairman of DGMH and editor of GSM 03.40[2] my first responsibilities in an international collaboration. In Telenor (at that time Televerket, Norway) I had worked in the research department on mobile communications since 1980. Much of this time was spent on mobile messaging, mainly in two business areas:

- for professional users: mobile message services for fleet management (dispatch companies, taxis, mail distribution, etc.);
- for both private and professional users: studies of how message handling systems and services could be accessed by mobile users.

In the mid-1980s I was project manager for a set of research activities that encompassed:

- market analysis of the above two areas of services;
- analysis and design of a messaging platform based upon X.25 that would combine both types of service.

The market analysis showed – not surprisingly – that the Nordic market was far too small for the provision of services based upon a system like this. However, the work with the project was in many respects of great value, e.g. when it came to assessing the potential of mobile messaging for a bigger market. It was with the experiences from those activities that I entered the work in DGMH.

The instructions that IDEG were given for the provision of messaging services fitted quite well with the experience gained from participation in the activities described above. There were several reasons for this, some very obvious. My superiors in the research department were also taking part in the GSM discussions that yielded the instructions! Also, our research department did not have the thick walls that separated the radio and data communication activities of the operating divisions in many telecom companies at that time. New concepts that combined the two areas therefore did not need the same time to

---

[1] As defined in IDEG doc. 7/87, rec. GSM 02.03 'Teleservices'.

[2] An overview of the evolution of the recommendation GSM 03.40 is given in Annex 7.

mature. And the services called for within the scope of DGMH were indeed new – in contrast to most of the other GSM services, most of which were ISDN replicas in a mobile context.

Anyway, I think it is fair to say that the newborn IDEG possessed quite vague notions of what the contour of the messaging services of GSM should be like and – even more – which market potential the services represented.

4.3   The Instructions that IDEG Were Given for Provision of SMS

In the pile of input documents to the first meeting, the following referred to SMS:

- the input document from WP1 on services;[3]
- a document from France envisaging solutions to some of the design challenges of SMS.[4]

The first document envisaged three categories of short message services:

- SM-MT/PP (Short Message Mobile-Terminated Point-to-Point) – the GSM service by which a short message might be transferred to a mobile station (MS).
- SM-MO/PP (Short Message Mobile-Originated Point-to-Point) – the GSM service by which a short message might be transferred from a mobile station (MS).
- SM-CB (Short Message Cell Broadcast) – the GSM service by which a short message might be transferred to a set of mobile users on a broadcast basis.

For both SM-MT/PP and SM-MO/PP, two very essential requirements applied:

- An SMS-SC (Service Centre) should act as a store-and-forward node for the messaging process.
- A short message should be transferred to or from the mobile station both when idle or in a state of call-in-progress. This implied that the transfer had to rely upon signalling resources, not traffic channels dedicated for transfer of user information on a circuit-switched basis.

The French document envisaged an overall procedure by which a short message might be transferred from the SMS-SC to the MS, indicating several scenarios of transfer through the GSM PLMN and including acknowledgement facilities. The procedure was based upon much of the procedure of a circuit-switched call set-up, indicating that much of the signalling resources established for this process might be applicable to SM-MT/PP. The French document proved to be a very useful document, enabling the DGMH crew to get on the right track faster than it otherwise might have done.

---

[3] IDEG doc 7/87, rec. GSM 02.03 'Teleservices'.
[4] IDEG doc 16/87, draft rec. GSM 10.02 'Technical Realisation of Short Message MT/PP Service'. (The GSM specification number is probably a misprint; it obviously refers to the planned GSM 03.40.)

Even though the two documents referred to represented substantial and sufficient guidelines for starting the design work on SMS, the start-up time for the design of SMS seemed to take somewhat longer than the design of the other services for which IDEG was responsible. The obvious reason for this was that SMS was the only one not representing a replica of an already existing service within ISDN or another fixed network. SMS thus required more time for consideration of market aspects than did the other services.

4.4 Overall Description of the Work in the Period from 1987 to 1990 and Work Items Dealt with

4.4.1 General

Through the first three years of DGMH, it is fair to say that the basic design was established. Major extensions and substantial enhancements and corrections were submitted as change requests after September 1990. However, the supporting walls and basic elements of SMS were all in place by mid-1990.

Twenty-one DGMH meetings were held during those three years. Seventeen of those were meetings being conducted in parallel with the meetings of IDEG/WP4/GSM4, and the other four were meetings of DGMH alone.

Forty-six different experts attended one or more of those meetings. However, only about eight people attended approximately one-third or more of the meetings.

Of the approximately 280 documents in the document stream of IDEG/WP4/GSM4 in the first three years that related to DGMH, approximately 120 were SMS-specific documents that DGMH either dealt with or produced. Approximately 30 DGMH relevant documents either dealt with or produced were on non-SMS issues (cell broadcast, MHS, etc.).

Most of the record of DGMH's first three years, as described below, is contained in the DGMH meeting reports to WP4/GSM4.[5] In two GSM4 meetings – Padua and Bath, spring 1990 – the DGMH reports were integrated into the GSM4 meeting report.[6]

4.4.2 Network Architecture and Transport Mechanisms

One of the first work items to deal with was the stabilisation of the network topology. The need for a store-and-forward node – the SMS-SC – was confirmed, as was the decision that this node should reside outside the PLMN.

Further, the requirement of the mobile user being able to send and receive short messages during a call in progress confirmed the already existing assumption that the signalling resources in such situations had to be used for the message transfer. As substantially different

---

[5] WP4 documents 71/87, 126/87, 151/87, 32/88, 87/88, 116/88, 148/88, 165/88, 208/88, 251/88, 298/88, 363/88, 16/89, 58/89, 65/89, 72/89, 116/89, 153/89 and 128/90.
[6] WP4 documents 42/90 and 77/90.

transfer schemes for the idle state and the call-in-progress state probably would require quite complex functionality, it was decided to base the message transfer upon the signalling capabilities for both situations.

In parallel with the discussions over the topics outlined above, the network topology emerged for transfer of short messages (see Figure 4.3).

Owing to the decision above, there were obvious similarities between the transfer of short messages and the set-up procedures for a speech or data call. In both cases the goal was to establish the necessary signalling capabilities over the radio path to or from the mobile station. It was therefore only natural that the relevant specification (GSM 04.11, see Section 4.8.1) had the same basic functional elements as the signalling capabilities for setting up a regular call (GSM 04.08).

Whereas the choice of transport capabilities was pretty obvious, the corresponding choice of transport capabilities in the core network – i.e. between MSCs – was not. Considerable time and effort were consumed on that issue. A true review of those discussions cannot be achieved without taking into consideration the strong standing that the X.25-based protocol stack of OSI possessed in those days. Telecom people throughout Europe and also other expert communities held a firm belief that the OSI protocols would constitute the bandwagon of data communications in the future. Therefore, there were several voices heard that advocated X.25-based OSI as a basis for the protocols both on the MSC ↔ MSC interface and on the SMS-SC ↔ MSC interface.

The following characteristics of the MSC were important. It was well established within the Signalling System No. 7-based MAP environment, and it did not beforehand have any X.25-based applications – either for signalling or for handling of user information. For these reasons it was decided, after some discussion, to develop the required functionality for transfer of short messages within the core network (MSC ↔ MSC) based on Signalling System No. 7 rather than X.25 (see also Sections 4.6.2 and 4.8.2).

However, for the SMS-SC the situation was quite different. The SMS-SC was like a blank sheet of paper – there were no prerequisites to be considered when entering the design phase. Furthermore, it had been decided to locate the SMS-SC outside the scope of PLMN. Bearing these aspects in mind, it was decided to specify an optional protocol based upon X.25 for the SMS-SC ↔ MSC interface (see also Section 4.6.3).

For both alternatives for transfer of short messages between a pair of network nodes, the application protocol ROSE (Remote Operations Service Elements) would apply. ROSE was originally defined by the OSI community to run on OSI protocols such as X.25, but had also been defined as a part of Signalling System No. 7.

After the topology had been assessed, work started on defining in more detail the transfer of short messages to and from the mobile station. The outcome converged towards the procedures depicted in Figures 4.5 and 4.6. In addition, work was carried out to define any possible error situation and to define how this should be reported to the sending party.

The issues mentioned above constituted a major part of DGMH's work on SMS during the first year of WP4's existence.

In addition to DGMH meeting reports, network architecture and transport mechanisms were addressed in more than 40 WP4/GSM4 documents.[7]

4.4.3 Service Elements

In parallel with the above activities, the work started to establish a service definition and to elaborate which service elements should be incorporated in the SMS.

It was decided to restrict the basic service to the transfer of short messages in both directions and the returning of the corresponding confirmations or error indications. At this first stage there was no proposal to elaborate procedures to be encompassed by the short message recommendation GSM 03.40 for SMS-SC ↔ MS conversation only, e.g. for address conversion in the SMS-SC (see 4.6.6). Thoughts on the feasibility of such procedures might have lingered in DGMH even in those first three years, but they were probably diverted by the assumption that this might be dealt with in higher protocol layers and by bodies outside the GSM community.

What should then be the service elements of the short message to be transferred?

The philosophy of DGMH when elaborating this area was to some extent influenced by work on message handling in the fixed networks – e.g. the X.400-based MHS concept of ITU and ISO.

For any message handling system that will make frequent use of a store-and-forward functionality, a time-stamping mechanism is required. The receiving party would often like to check when the sending party submitted the message for transfer. This was considered to be especially important for mobile communications systems, where mobile stations were expected to be turned off or out of coverage for longer periods of time. The Service-Centre-Time-Stamp protocol element was therefore incorporated in the message header (see also Section 4.5.4).

Again owing to the expectation that mobile stations might be unreachable for substantial periods of time, the Validity-Period parameter was introduced in the short message header. If a short message happened to be stored after the Validity-Period had expired in the SMS-SC, e.g. owing to the recipient being absent, the SMS-SC was allowed to delete the short message (see also Section 4.5.4).

Like many other protocols for which one might at some stage experience a need to carry an application protocol, it was considered beneficial to include a parameter that would identify *which* application protocol was being carried at the moment. A Protocol-Identifier parameter was thus included in the short message header, but without a specific set of values for a corresponding set of application protocols. The inclusion of the Protocol-Identifier was a typical example of just reserving space for an element that later on might prove to be useful (see also Section 4.5.4).

---

[7] WP4 documents 54/88, 66/88, 85/88, 101/88, 106/88, 124/88, 125/88, 183/88, 228/88, 250/88, 262/88, 300/88, 301/88, 324/88, 332/88, 334/88, 339/88, 17/89, 18/89, 19/89, 73/89, 75/89, 76/89, 77/89, 80/89, 81/89, 84/89, 85/89, 94/89 and 105/89, GSM4 documents 156/89, 161/89, 185/89, 14/90, 19/90, 24/90, 74/90, 90/90, 115/90, 116/90, 117/90 and 125/90.

In view of the same assumption that a mobile station might be unreachable for a rather long period of time, the possibility was considered of a queue of short messages waiting in the SMS-SC to be transferred. There were early considerations of the aspects of speeding up message transfer at the lower layers by keeping signalling resources established instead of setting them up and releasing them each time they were needed (see also Section 5.2.1 on continuous message flow). For example, the first approved version of GSM 04.11 (see Section 4.8.1) was designed in such a way that the signalling resources were kept after a short message transfer attempt to handle the following confirmation or error indication. According to the ongoing prioritising of work items of DGMH, it was seen as a sensible strategy first to stabilise the design of the simple and less efficient structure of handling each short message transfer and its confirmation separately, and later to improve the efficiency of transfer of short messages queuing up in the SMS-SC, as indicated. However, already at this stage a specific parameter More-Messages-to-Send was included in the short message header (see Section 4.5.4). The value of the More-Messages-to-Send field in a short message to be transferred from the SMS-SC to the MS would indicate if there were more messages waiting in the SMS-SC to be transferred to that MS. The parameter would come into use, for example, when the signalling resources were to handle a sequence of short message transfers instead of just one.

When a mobile station that has been unreachable for a while turns up again, there might be messages waiting in the SMS-SC to be transferred to it. What should the procedures be for retrieving the short messages in such instances? Should the SMS-SC apply some periodic transfer attempt procedure, or should mobile users themselves take action by some manual means when they detect that they are once again in contact with the mobile network? Schemes like this seem neither elegant nor efficient. A preferable solution would imply some mechanism that transferred the waiting messages:

- as soon as the mobile station is able to receive them and without any manual action required from the mobile user;
- without any scheme of numerous and superfluous transfer attempts from the SMS-SC's side.

The solution that was chosen consisted of functional elements to be invoked under two different situations:

1. When the SMS-SC fails to deliver a short message to a mobile user owing to the user's mobile station being switched off or being out of coverage. In this situation, a flag (Messages-Waiting-Flag) is set, and the address of the SMS-SC is stored in the location register of the mobile station (Messages-Waiting-Data). The service element that enabled the network to set the flag and store the address was denoted Messages-Waiting.
2. When a location register with the above flag set detects that the mobile station is again both turned on and connected to the network. In this situation, the network sends an alert to each of the SMS-SCs that have tried to transfer short messages to the mobile station while it was not reachable. The SMS-SC will then make another attempt to transfer the

waiting messages. The service element that enabled the network to alert the appropriate SMS-SCs in that situation and thereafter remove the relevant flag and addresses was denoted Alert-SC.

The service element Messages-Waiting was very much related to a paired functionality introduced earlier by WP3, namely IMSI Attach and IMSI Detach. IMSI Detach was a procedure invoked by the mobile station when the mobile user switched it off. Before actually turning the power off, the mobile station would send a signalling message to the network, asking the location register to set an IMSI Detach flag. Correspondingly, when the mobile user switched on the mobile station again, a signalling message would automatically be sent to the network with a request that the IMSI Detach flag be removed. The intended feasibility of IMSI Attach and IMSI Detach was that they would spare the network from superfluous signalling efforts, e.g. from incoming calls to mobiles being switched off. The functionality of IMSI Attach and IMSI Detach was optional, i.e. PLMN operators did not have to implement them. At the beginning of the GSM era, some operators were sceptical owing to the fear that a tidal wave of almost simultaneous IMSI Attach messages, e.g. in the morning, would jam the signalling capabilities of the GSM network. As this fear proved to be somewhat exaggerated, IMSI Attach and ISMI Detach were implemented in almost every GSM network.

An IMSI Detach flag set in the location register would simplify a short message transfer attempt in the sense that it would cancel the attempt, set the Messages-Waiting-Flag (if not set already) and see to it that the SMS-SC address was contained in the Messages-Waiting-Data in the location register. In a substantial part of the cases where the Messages-Waiting service element would come into operation, it would be invoked by a previous IMSI Detach.

In almost every discussion about designing new telecom services at that time, different aspects of priority became an issue. This was also the case with the design of SMS. Some viewpoints predicted requests from authorities for a way to give certain short messages priority above 'regular ones'. DGMH gave those aspects thorough consideration, but did not find many obvious processes in the transfer of short messages in which priority settings would have much effect. In situations where messages were queuing up in the SMS-SC waiting for the mobile station to become reachable again, one could of course modify the sequence – putting the priority messages first in line. This would not matter much, as it was expected that the transfer of each message would not take long. The last message in a queue of four or five messages would only come about 30 seconds later than the first one. DGMH therefore decided to concentrate the Priority service element at another stage of short message transfer: if a message with the Priority parameter set encounters the location register of the mobile station with the Messages-Waiting-Flag set, the network will not suspend the transfer attempt as it would do in the case of a non-priority message – it will try to deliver the message in spite of the flag being set.

The reason for this is that there is a marginal probability that the flag may not be removed when it should. If the mobile station is out of coverage for a short time – e.g. mounted in a car driving through a tunnel – and then re-enters coverage in the same cell, that last event will not be notified by the location register. If there has been an attempt to transfer a short message during the period of time in which it has been absent and the flag has been set, it may stay set for a while even after having re-entered coverage, i.e. until there is some signalling procedure that involves both the mobile station and the location register.

Now, DGMH decided that the Priority service element should relate to the above possibility: if there was a transfer attempt of a short message with the Priority parameter set, the transfer attempt would be carried out even if the flag was set. Priority in this context thus meant 'guaranteed delivery if at all possible'.

The service aspects of SMS as depicted above and in Section 4.4.4 represent some of the major topics to be dealt with from summer 1988 until spring 1990.

In addition to the DGMH meeting reports, the service elements of SMS were addressed in 13 WP4/GSM 4 documents.[8]

4.4.4 Supplementary Services

DGMH also considered whether it was feasible to align SMS with ISDN in terms of supplementary services. For many supplementary services of ISDN, no parallel for SMS could be established, such as, for example, user-to-user signalling. For others, no parallel appeared to be suitable, such as, for example, completion of calls to busy subscriber. Most time was spent on a family of supplementary services that at first glance seemed to be suitable also for SMS: the call forwarding services. Again, for some of those an adaptation to SMS was hardly feasible. For instance, call forwarding on busy subscriber seemed to be pretty useless because a short message could be delivered to the mobile user even if there was a call in progress. Call forwarding on no reply or call forwarding on absent subscriber also seemed to be of little interest – the store-and-forward functionality was intended to compensate for recipients being temporarily unable to receive the information.

Call forwarding unconditional was different, however. There might be several good reasons for including that service in the repertoire of SMS. When starting to explore a possible design for SMS call forwarding unconditional, a set of technical questions arose. It was obvious that the interface between the SMS-SC and the PLMN would have to be far more complex if that supplementary service were to be incorporated. Also, some more basic considerations were made: call forwarding unconditional – which does not incorporate a warning to the sender that the short message will be received by somebody other than the intended recipient – may be less suitable in view of some code of conduct than the call forwarding unconditional for regular voice calls. For voice forwarded calls, the caller gets in direct contact with the user to which the call has been forwarded, and the call may be moderated accordingly. In messaging, this is not the case. The user to which the message has been forwarded will receive the text that the sender intended for that recipient only. In email communication, forwarding is quite frequently used, but almost only in a manual and individual way; i.e. the intended recipient may – after having scrutinised the incoming mail and considered its interest to others – take manual action to forward it.

---

[8] WP4 documents 113/88, 319/88, 333/88, 354/88, 358/88, 359/88, 50/89, 77/89 and 80/89, GSM4 documents 130/89, 25/90, 31/90 and 33/90.

Having discussed aspects of the topics mentioned above, DGMH decided to recommend that no supplementary service be incorporated at this stage in the SMS. The most important reason for this decision was the wish to keep SMS as a very simple and non-complex service.

In addition to the DGMH meeting reports, the supplementary services were addressed in three WP4/GSM4 documents.[9]

4.4.5   Short Message Length and Alphabet

As mentioned above, SMS was from the early phase of conception intended to differ from the circuit-switched services of GSM in many respects, and also in that it would have to be carried by the signalling capabilities over the radio path. This conclusion came from some of the basic service requirements, e.g. the requirement that short messages should be sent or received during a circuit-switched call in progress. This reinforced the perception of a short message as being a small piece of text. The earliest preliminary assessments of the maximum length of short messages were 256 and, later, 180 bytes.

It gradually became apparent that the signalling capabilities on the leg between the mobile station and the MSC did not comprise any functionality that contained a hard limit with respect to length. Owing to the fact that the radio transmission was based upon protocol elements of relatively short blocks, it was evident that, even for transfer of relatively small amounts of data, the corresponding protocol would have to cope with sending several blocks in sequence. Hence, the radio transmission was not the limiting factor. However, when the decision was taken to use MAP for transfer between MSCs, it was obvious that a similar investigation had to be conducted for the specific MAP operation. As outlined in Section 4.5.8, it appeared that, owing to the use of MAP, one had to restrict the maximum length of short messages to 140 bytes instead of 180.

The reduction in the maximum message length did not alarm the DGMH or GSM4, but it certainly focused attention on how to exploit the message volume in the best way. Would it be possible to reduce the number of bits per character and still be able to establish an alphabet that would fulfil the requirements from SMS users? As powerful messaging systems emerged – such as the X.400 standard of MHS by ITU and ISO – a great interest in standardisation of feasible alphabets evolved. Such work items were also the natural responsibilities of ITU and ISO. The most common standard in the 1980s was IA5 (International Alphabet No. 5) defined in the ITU recommendation T.50. IA5 of T.50, as defined in 1988, was a 7-bit character alphabet based on the Latin character set but subject to a great variety of extensions and modifications. Those would, however, only be achievable in ways that were not easily handled by SMS. As DGMH was looking for an alphabet based on the most frequently used characters in Europe at that time, IA5 was apparently not an optimal choice.

At that time, DGMH became aware of the work that been done in the ERMES groups, which were assigned to specify a text paging system for Europe. ERMES had aimed at more

---

[9] WP4 document 83/89, GSM4 documents 131/89 and 132/89.

or less the same goal as DGMH with SMS: an alphabet of 7-bit characters that comprised as many as possible of the most frequently used characters in Europe. The result was, of course, a compromise in which some of the less frequent characters in the smaller languages were missing (see also Section 4.5.9 and Figure 4.8). However, DGMH considered that this was as close as one could come, and recommended that the ERMES alphabet replace the IA5 alphabet in GSM 03.40 that had been adopted on a preliminary basis. GSM4 endorsed the recommendation. Eventually, the ERMES alphabet, along with a maximum length of short messages of 160 characters, was approved by GSM1.

The issues of maximum short message length and the alphabet of SMS were dealt with between autumn 1989 and spring 1990. In addition to the DGMH reports, seven WP4 documents dealt with the problems of alphabet and short message length.[10]

4.5 The SMS of September 1990

4.5.1 Some General Remarks on the Way that Services, Networks and Communication Protocols Were Described in the Late 1980s

Design of communication systems in the 1980s had the benefit of relying on a detailed and well-defined formalism. For connection-oriented services in particular, endeavours to define a strict model for the structure of service definition, communication protocols and protocol layering had resulted in a set of working procedures that made it easy to gather experts from different parts of the world to join in and contribute to a broad collaboration like GSM. It also helped that the development of GSM was conducted partly just after and partly in parallel with the development of ISDN.

Owing to the fact that a substantial part of the work on the design of SMS during the years 1987–1990 was devoted to the challenge of adapting SMS to the network topology and protocol structure of the circuit-switched domain of the GSM PLMN, this section offers a quick review of the basics of protocol structuring in the OSI context.

Some of the core principles of service definition, communication protocols and protocol layering are encapsulated in Figure 4.1.

In Figure 4.1, protocol elements (or, in OSI terminology, protocol data units (PDUs)) at protocol layer n may be sent between entities A and C. A and C are not directly connected, only via node b at a lower protocol layer.

When sending a protocol message from A to C, A's entity at layer n instructs A's entity at layer $n - 1$ to take action to make this happen. In OSI terminology, A's entity at layer n invokes the appropriate service primitive with the relevant parameter values of the PDU to be sent at layer n (green arrow pointing downwards in Figure 4.1). A's entity at layer $n - 1$ establishes one or more PDUs to be sent at this layer to b. As indicated in Figure 4.1, PDUs may be sent in both directions for both links A $\leftrightarrow$ b and b $\leftrightarrow$ C; e.g. acknowledgements may be sent on the paths b $\rightarrow$ A and C $\rightarrow$ b.

---

[10] WP4 documents 102/88, 108/88, 78/89, 86/89 and 103/89, GSM4 documents 177/89 and 22/90.

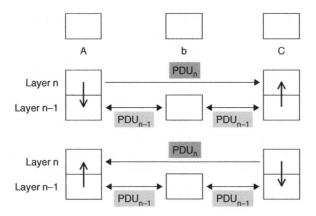

Figure 4.1 Protocol layers n and $n-1$ connecting the three nodes A, b and C. A set of available PDUs are defined between each pair of nodes at the different protocol layers and constitute the respective protocols. The vertical arrows in green represent the service primitives that are used to send PDU_n between A and C

A PDU at any layer comprises two parts: a header with protocol parameters, address information at that level, etc., and a user information part (see Figure 4.2). At the uppermost layer level, the user information part is the core information that the sending party wants to convey to the receiving party. Going down the layers, the different entities at the sending node add ever more data to the user information by including the header at the layer above. The physical transfer of the core information – or the user information at the upper layer – may thus get a substantial overhead, depending upon how many layers the protocol hierarchy consists of and how much header information is required at each layer.

Sending a protocol message from A to C in Figure 4.1 is then achieved in the following way: A's entity at layer $n-1$ receives a service primitive from layer n, which results in PDUs to be transferred between A and b and between b and C at layer $n-1$ to get the core information to node C. It is presented to the entity at layer n in that node by a service primitive (arrow pointing upwards in node C).

The repertoire of PDUs available between two nodes and the rules for how they should be used is said to constitute the protocol between the two nodes at that layer.

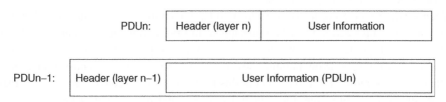

Figure 4.2 A PDU at any protocol layer n is divided into the header and the user information part. Taking the two layers n and $n-1$, the user information at layer $n-1$ will equal the PDU at layer n

The set of service primitives available between two layers at a certain node is said to be the definition of the service that the lower layer is offering to the layer above it at that part of the network.

For communication networks the OSI standard of ITU and ISO defined a general protocol hierarchy of seven possible layers, in which each layer served a particular purpose. Four of the OSI layers ought to be mentioned here: the link layer (layer 2), the network layer (layer 3), the transport layer (layer 4) and the application layer (layer 7).

The link layer would cater for transfer of data over the link between a pair of nodes.

The network layer[11] would deal with address information and routing in such a way that the layer above – the transport layer[12] – and the layers above that would be relieved of this task.

The application layer (layer 7) would encompass any application protocol to be implemented between any two nodes.

The OSI structure constituted the basis upon which all communication protocols within GSM system were based. As mentioned above, the OSI model proved to be a very powerful and versatile concept in the design of communication systems. However, it also represented some minor challenges. One of those became apparent when a protocol hierarchy with layers from 1 to 7 – such as MAP based upon Signalling System No. 7 – was to be the carrier of another type of information, i.e. when one placed protocols for transport of short messages on top of MAP (see also Section 4.5.5). It might introduce some confusion in terms of protocol architecture, and it would result in a significant overhead.

4.5.2 The Specifications Defining the Short Message Service Mid-1990

The short message service was in mid-1990 defined by recommendations GSM 03.40 v3.4.0, GSM 04.11 v3.1.1 and GSM 09.02 (MAP) v3.5.1. The evolution of GSM 03.40 is shown in the table in Annex 7.

4.5.3 Network Architecture

The network architecture defined in the 1990 versions of GSM 03.40 is given in Figure 4.3.

---

[11] From Wikipedia, 24 July 2009: The Network Layer is responsible for end-to-end (source to destination) packet delivery including routing through intermediate hosts, whereas the Data Link Layer is responsible for node-to-node (hop-to-hop) frame delivery on the same link. The Network Layer provides the functional and procedural means of transferring variable length data sequences from a source to a destination host via one or more networks while maintaining the quality of service and error control functions.

[12] From Wikipedia, 24 July 2009: In computer networking, the Transport Layer is a group of methods and protocols within a layered architecture of network components within which it is responsible for encapsulating application data blocks into data units (datagrams, segments) suitable for transfer to the network infrastructure for transmission to the destination host, or managing the reverse transaction by abstracting network datagrams and delivering their payload to an application. Thus the protocols of the Transport Layer establish a direct, virtual host-to-host communications transport medium for applications and are therefore also referred to as *transport protocols*.

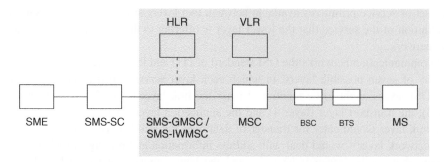

Figure 4.3 The network architecture of SMS according to the 1990 versions of GSM 03.40. The grey coloured zone defines the PLMN area. The links on which the short messages are conveyed are given by the unbroken lines. The links on which only signalling information is conveyed are given by the dotted lines. The BSC and BTS are shown in smaller size and fonts because the protocol layer defined in GSM 03.40 does not involve these nodes

The different nodes of the SMS network architecture were as follows:

- SME, Short Message Entity. Any entity capable of sending or receiving short messages to or from a mobile station. The most obvious alternative was, of course, a GSM mobile station in a GSM PLMN, but it could also be, for example, a Telex terminal in the Telex network, a telephone with alphanumeric keys in the PSTN or a PAD terminal connected to the PSPDN. The SME potential was completely dependent upon the interworking capabilities of the SMS-SC.
- SMS-SC, Short Message Service Centre. A node that takes care of the store-and-forward functionality of messages to and from the mobile station.
- SMS-GMSC/SMS-IWMSC. An MSC that is connected to the SMS-SC. In view of the fact that such an MSC will have slightly different functions in the mobile-terminated and mobile-originated cases, it is denoted SMS-GMSC and SMS-IWMSC respectively.
- MSC (optional). An intermediate MSC between the SMS-GMSC/SMS-IWMSC and the MS will be required when the MS does not reside within the coverage area of any BSC/BTS connected to the SMS-GMSC/SMS-IWMSC.
- MS, Mobile Station.
- HLR, Home Location Register. The part of the subscriber data that is always located in the home network of the user.
- VLR, Visitor Location Register. The part of the subscriber data that is connected to the current state and whereabouts of the user. The VLR is dynamic and fluctuant in the sense that its data are valid only when the mobile station is switched on and has been authenticated, and that the register is physically established in the part of the network where the mobile station currently is.

In the following, most references to the data of the HLR or VLR are given in a context where it is not necessary to distinguish between the two. In those cases, the term 'location register' is used as a common term.

As for the specification for most network nodes of the GSM PLMN, apart from overall functionality requirements, they have detailed specifications of the interfaces between them. This is also the case with SMS.

According to the overall requirements of the SMS-SC, it should be capable of:

- submitting a short message to an MS and retaining the responsibility for the message until the report has been received or the Validity-Period expires;
- receiving a report from the PLMN;
- receiving a short message from an MS;
- returning a report to the PLMN for a previously received short message.

4.5.4 Service and Service Elements

According to GSM 03.40 of September 1990, the SMS comprised seven elements particular to the submission and reception of messages:

- Validity-Period;
- Service-Centre-Time-Stamp;
- Protocol-Identifier;
- More-Messages-to-Send;
- Priority;
- Messages-Waiting;
- Alert-SC.

The work with and development of the service and service elements of SMS are described in Section 4.4.3.

The Validity-Period information element indicates the time period for which the short message is valid, i.e. the length of time that the SMS-SC should guarantee its existence in the SMS-SC memory before delivery to the recipient has been carried out.

The Service-Centre-Time-Stamp information element allows the SMS-SC to inform the recipient MS about the time of arrival of the short message at the SMS-SC.

The Protocol-Identifier information element allows the SM-TL to refer to the higher-layer protocol being used, or indicate interworking with a certain type of telematic device.

The More-Messages-to-Send information element allows the SMS-SC to inform the MS that there are one or more messages waiting in the SMS-SC to be delivered to the MS.

The Messages-Waiting information element enables the PLMN to inform the HLR and VLR (with which the recipient MS is associated) that there is a message in the originating SMS-SC waiting to be delivered to the MS. The information element is only used in cases of previous unsuccessful delivery attempt(s) due to a temporarily absent mobile. This information, denoted the Messages-Waiting-Indication (MWI), consists of Messages-Waiting-Data (MWD) located in the HLR and the Messages-Waiting-Flag (MWF) located in the VLR. The MWD contains a list of addresses of SMS-SCs that have made previous unsuccessful delivery attempts of a message. In order to link an alert message to an earlier delivery attempt in the multiple MSISDN scenario, the HLR will store each SMS-SC address.

The Priority information element is provided by an SMS-SC or SME to indicate to the PLMN whether a message is a priority message. Delivery of a non-priority message will not be attempted if the MS has been identified as temporarily absent. A priority message will be given a delivery attempt irrespective of whether the MS has been identified as temporarily absent or not.

There are two ways in which the mobile station might be notified by the network of being temporarily absent:

- The mobile station declares itself absent by sending an IMSI Detach by an action included in the internal procedures when being turned off. The opposite action would be to send an IMSI Attach by an action included in the internal procedures when being turned on.
- There would be one or more unsuccessful short message transfer attempts from the SMS-SC to the mobile station, resulting in the MWF being set (see above).

The Priority information element does not involve prioritisation of the message in any queueing or handling in the SMS-SC or MSC or by MAP.

The Alert-SC service element may be provided by some GSM PLMNs to inform the SMS-SC that an MS to which there has been an unsuccessful delivery attempt and which is now recognised by the PLMN as having recovered operation (e.g. as having responded to a paging request) is again attainable. The SMS-SC may, on reception of an Alert-SC, initiate the delivery attempt procedure for the queued messages destined for this MS.

4.5.5   Protocol Architecture and Service Definition

The protocol architecture of SMS is given in Figure 4.4.

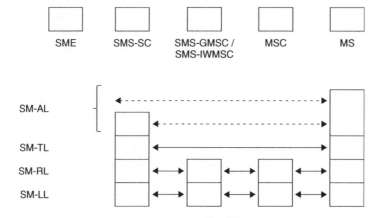

Figure 4.4   Protocol architecture of the SMS

The protocol layers defined for SMS were:

- SM-TL (Short Message Transfer Layer). This was the layer containing the protocol defined by GSM 03.40.
- SM-RL (Short Message Relay Layer). This was the layer containing the upper-layer protocols of the interfaces between the involved nodes: (i) SMS-SC ↔ SMS-GMSC/SMS-IWMSC; (ii) SMS-GMSC/SMS-IWMSC ↔ MSC; (iii) MSC ↔ MS. On interface (i) the optional protocol of remote operations based upon X.25 was offered. However, the SMS-SC and the PLMN operator could agree to use an alternative protocol. On interface (ii), a MAP operation applied. On interface (iii), the GSM 04.11 applied.
- SM-LL (Short Message Lower Layers). These were layers for all protocols at levels below SM-RL.

In addition, SM-AL (Short Message Application Layer(s)) are depicted. These were the protocol layers outside the scope of ETSI.

As mentioned in Section 4.5.1, the decision to base the transfer of short messages between MSCs upon MAP makes all layers SM-AL, SM-TL and SM-RL reside in the OSI application layer.

4.5.6   Principle Schemes of Short Message Transfer

4.5.6.1   The Mobile-terminated Case

In Figure 4.5 the successful case of a mobile-terminated transfer is shown.

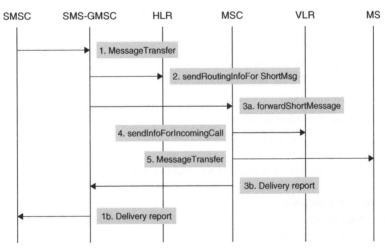

Figure 4.5   A mobile-terminated short message transfer – the successful case

4.5.6.2   The Mobile-originated Case

In Figure 4.6 the successful case of a mobile-originated transfer is shown.

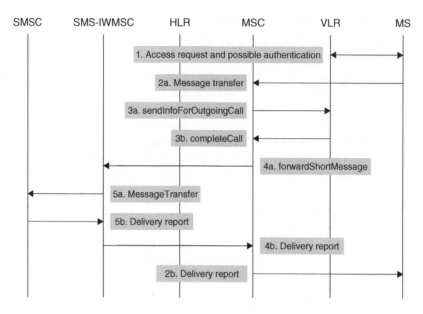

Figure 4.6 A mobile-originated short message transfer – the successful case

4.5.7 Addressing Capabilities

The addresses of the PDUs at the SM-TL and SM-RL would consist of the following elements:

- an Address-Length field of 1 byte;
- a Type-of-Address field of 1 byte, containing
    (i) a Type-of-Number and
    (ii) a Numbering-Plan-Identification;
- an Address-Value field of variable length.

The Type-of-Number would indicate if the address was either:

 (i) an international number;
 (ii) a national number;
(iii) a network specific number; or
(iv) a short number.

The Numbering-Plan-Identification would indicate if the number should refer to (i) the ISDN/telephone numbering plan (CCITT rec. E.164/E.163), (ii) the data numbering plan (CCITT rec. X.121), (iii) the Telex numbering plan, (iv) a national numbering plan or (v) a private numbering plan.

4.5.8 Maximum Length of Message

It was clear from the very conception of the short message service that the design of the service should be made under the assumption that there would be no transfer of long messages. If there was a need for such services, they should be catered for by other capabilities of the GSM system.

But what was the maximum length of messages that *could* be transferred by SMS?

It became clear that the limiting functional element was the MAP operation forwardShort-Message (see Figures 4.5 and 4.6) for conveying messages between MSCs. As for every MAP operation, this was based upon the use of TCAP (Transaction Capabilities Application Part). TCAP was defined in ITU-T recommendations Q.771 to Q.775 or ANSI T1.114 as a protocol for Signalling System No. 7 networks. Its primary purpose was to facilitate multiple concurrent dialogues between the same subsystems on the same machines, using transaction IDs to differentiate these, similar to the way TCP ports facilitate multiplexing connections between the same IP addresses on the Internet.

When choosing the simplest scheme of TCAP as a basis for the MAP operation forwardShortMessage, one would find the maximum number of bytes available for the user information field of the short message by considering (i) the gross number of bytes available and (ii) the overhead at the different layers of the protocol stack of the signalling system.

The first calculations of the overhead budget, from the lowest layers of Signalling System No. 7 to MAP at the top, indicated that slightly more than 140 bytes would be available for genuine message information. Then, 140 bytes – or 160 7-bit characters – was established as the 'official' value of the short message length. In Figure 4.7 an extract is given from the

REASON FOR CHANGE: Defining the SM length for SM service.
The maximum short message length is limited by the overhead of the MAP operation "Forward short message", which leaves 171 octets for the TPDU. The maximum short message length is 146 octets, if the originating address in the SMS-SUBMIT (which is anyhow conveyed at the relay layer) is omitted.
For practical reasons, the maximum short message length is set to 140 octets.
However, the originating address must be provided by the SM-RL.

DETAILS OF CHANGE:
Omitting the SMS-SUBMIT field TP-OA on the pages 45 and 46.
Modifying the RS-MO Data definition §9.3.3.2 page 57.
Modifying the RP-MO Data definition §9.3.5.1 page 61.

IMPACT ON OTHER RECOMMENDATIONS: Within GSM 02.03 Annex to datasheet of teleservice 21 to 23, replace the sentence "the messages are limited to a length of 180 octets" with "the messages are limited to a length of 140 octets".

Figure 4.7 Extract from the change request to change the short message length[13]

---

[13] WP4 document 86/89 rev. 1 'Change request to GSM 03.40: Short message length'.

change request submitted by DGMH to GSM4 in late 1989 to change the first assessment of 180 bytes to 140. GSM4 approved the request and liaised with GSM1 also to update GSM 02.03.

As the overhead budget in MAP and Signalling System No. 7 was subject to further consideration, e.g. because of the need to add even further protocol elements to the TPDUs, the exercise was also carried out after September 1990.[14] However, the review of the overhead budget did at that time alter the value of maximum message length to 140 characters.

4.5.9   Alphabet Available for User Information

With the limitation on the maximum length of a short message defined in the process indicated in Section 4.5.7, it was obvious that one had to look very carefully at the candidates of the alphabet to be used. Any requirements to allow for a large character set – encompassing most of the world's main alphabets – would increase the number of bits per character compared with the number needed for a narrow character set. The result would be a decrease in the maximum number of characters per short message. The goal was therefore to balance the need for the most significant characters in the written languages of Europe against the need to keep the character repertoire down. The best compromise seemed to be the choice made by ERMES for the 7-bit character alphabet intended for the new European paging service.

The character set recommended for SMS in 1990 is given in Figure 4.8.

4.6   Major Design Issues

4.6.1   General

During the period from May 1987 to September 1990, the four most crucial questions related to the design of SMS were:

- What transport service should be chosen to carry the short messages in the core network?
- As regards network topology, should the SMS-SC reside inside or outside the GSM PLMN?
- What interface should be chosen between the SMS-SC and the SMS-GMSC/SMS-IWMSC?
- As regards the topology of SMS-SC and MSC connections, what restrictions should apply?

To none of the four questions there was an intuitive answer, and different opinions about some of them remained for some time. For all of them the decisions that were taken had a crucial impact upon the feasibility of SMS.

---

[14] See, for example, Annex 5 of GSM4 document 126/91.

b4	b3	b2	b1	b7 0 / b6 0 / b5 0 / 0	b7 0 / b6 0 / b5 1 / 1	b7 0 / b6 1 / b5 0 / 2	b7 0 / b6 1 / b5 1 / 3	b7 1 / b6 0 / b5 0 / 4	b7 1 / b6 0 / b5 1 / 5	b7 1 / b6 1 / b5 0 / 6	b7 1 / b6 1 / b5 1 / 7
0	0	0	0	0	@	Δ	SP	0	¡	P	¿	p
0	0	0	1	1	£	1)	!	1	A	Q	a	q
0	0	1	0	2	$	Φ	"	2	B	R	b	r
0	0	1	1	3	¥	τ	#	3	C	S	c	s
0	1	0	0	4	è	λ	¤	4	D	T	d	t
0	1	0	1	5	é	Ω	%	5	E	U	e	u
0	1	1	0	6	ù	π	&	6	F	V	f	v
0	1	1	1	7	ì	Ψ	'	7	G	W	g	w
1	0	0	0	8	ò	Σ	(	8	H	X	h	x
1	0	0	1	9	Ç	Θ	)	9	I	Y	i	y
1	0	1	0	10	LF	≡	*	:	J	Z	j	z
1	0	1	1	11	Ø	1)	+	;	K	Ä	k	ä
1	1	0	0	12	ø	Æ	,	<	L	Ö	l	ö
1	1	0	1	13	CR	æ	-	=	M	Ñ	m	ñ
1	1	1	0	14	Å	β	.	>	N	Ü	n	ü
1	1	1	1	15	å	É	/	?	O	§	o	à

Figure 4.8 Table showing the character alphabet defined for SMS in 1990

4.6.2   Inter-MSC Transfer of Short Messages – by X.25 or Signalling Capabilities?

It was clear from the very start that, at least over the radio path, short messages had to be transferred by the signalling capabilities in order to allow a low-cost implementation in every mobile station and every base station. This was needed in order to fulfil the requirement that the reception or submission should be invoked even when a call (such as a speech call) was in progress. However, over some links within the core network, alternatives to the signalling system did exist.

The general path of transfer of short messages between the MS and the SMS-SC is depicted in Figure 4.3. The short message will pass one BTS and one BSC only, but may pass two MSCs. On the links from the BTS via the BSC to the MSC there was no economically reasonable alternative to using signalling channels of the GSM system. But on the MSC to MSC link, means other than the signalling capabilities were conceivable.

Several experienced experts did not approve of the principle of channelling user information – being either on a real-time or a best-effort basis – over the signalling system. One of the very basic rules of thumb from the POTS[15] era was that you should keep an idle lane on the signalling channel open as long as possible. If the signalling system was jammed, the users might be left without any service at all.

Thus, viewpoints were put forward to specify X.25 as the means of transfer of short messages between MSCs.[16] In this respect, X.25 was an obvious choice. Everything seemed to confirm that X.25 would be around for a while as one of the major transport mechanisms for data on a store-and-forward basis. It had a promising take-up in the market, and it was the fundamental principle upon which the ITU and ISO were building their OSI model.

Many supporters of using Signalling System No. 7 came from the new breed in the telecom business: the mobile operator – a company that only provided mobile services and was play-ing within a competitive arena. The old incumbents resided on several platforms – PSTN, circuit-switched data networks, packet-switched data networks and gradually PLMNs – thus being self-contained. Companies of the first category knew they would be faced with the situation where they had to purchase X.25 capacity from other network operators if they were to deploy SMS on such grounds. Signalling System No. 7, on the other hand, was 'free of charge' and within their own control.

After a discussion that spanned a couple of meetings, DGMH decided to rely on Signalling System No. 7 as the only alternative for the transfer of short messages between MSCs.

It is hard to overestimate the significance of this decision. As every GSM PLMN operator had to purchase Signalling System No. 7 as the basis of the Mobile Application Part sig-nalling system, an important hurdle was passed on the way to a regional or even global SMS network. One may wonder what would have been the result if X.25 had been a necessary add-on to GSM to cater for a small but required cogwheel for SMS. X.25 had its prime time

---

[15] POTS = Plain Old Telephones System/Service, a jargon widely used.
[16] See, for example, WP4 doc. 116/88.

in the 1980s and the early 1990s, but was more or less put aside as an alternative to a data network carrier when the Internet and IP technology conquered the world in the mid-1990s.

The solution of basing the transfer of short messages upon MAP and Signalling System No. 7 also represented some challenges. For instance, in later versions of MAP, the headers at lower layers were increased, thereby leaving less space for the message content (see also Section 5.2.14 on the upgrading of network capability and consequent reduction in SMS length). However, this more or less coincided with the steps that were taken by DGMH to increase maximum message length by concatenation, so the net change for the short message user was positive rather than negative.

4.6.3   Service Centre – Inside or Outside the PLMN?

From the very origin of the idea of including a text messaging service in GSM, it was obvious that a network node had to be introduced in addition to the PLMN-specific nodes for handling user traffic, i.e. the BTS, BSC and MSC. Many aspects led to this conclusion. In particular, there was a common understanding that the PLMN nodes as far as possible would be based on the technology of standard ISDN exchanges and would hence not contain any store-and-forward functionality. As messaging is tightly connected to the aspects of store-and-forward, one had to include another network node to take care of that. So the idea of a short message service centre was contemporary with the idea of a GSM short message service.

The standards of GSM were defined in a period of time when the telecom community became very much aware of the benefits of dividing the provision of network services and the provision of value-added services into different domains. There should be a well-defined and open interface between the two domains, so that players other than the network operator would have business opportunities based upon value-added services. Blurring the demarcation line between network services and value-added services by, for example, including elements of value-added services in the responsibility of the network operator would only hamper business development and possibly distort the balance of competition.

There was a strict discrimination between network services and value-added services: a functionality that in any way influenced or had an impact upon the bit stream of user information – like adding store-and-forward or changing the format of the PDUs – was considered to add value to the service and therefore to reside within the domain of value-added services.

Thus, from the very beginning the SMS-SC was located outside the PLMN according to the structure forced on network designers, as indicated above. However, when company contributions to the standards process contradicted this and included the SMS-SC within the PLMN as a regular GSM node, there was a discussion on this within the DGMH by reviewing both the principles and the practicalities of locating the SMS-SC. This discussion ended with the conclusion that the SMS-SC should be kept outside the PLMN.

In hindsight, it is an open question whether this was a wise decision.

Along with the attention and expectation of the new VAS domain, a working procedure emerged that one should define a demarcation line between value-added services and network services. Network operators should involve themselves as little as possible in the business and endeavours of VAS. In particular, they should not do so to gain an advantage in the competition in any of the two domains. A regulatory regime like this would yield a series of positive effects that would apply to SMS as well. It would pave the way for numerous VAS providers who would start off engineering a series of exciting new applications based upon SMS.

As we now know, this is not what happened. There were no crowds of VAS providers putting up their SMS-SCs to be connected to the recently launched PLMNs. To get a short message service going, mobile operators had to purchase and install the SMS-SC themselves; very few non-operators – if any – lined up to offer additional or competitive services. And when some eventually did, lengthy disputes between the PLMN operator and the operator of the stand-alone SMS-SC on interconnect and other regulatory issues might follow.

It should be concluded that, for the 20 years following the launch of SMS, nearly all of the SMS offerings in Europe stemmed from SMS-SCs that were owned and operated by PLMN operators. Some might also say that, for the sake of innovation of SMS and SMS applications, it might have been better to have left the SMS-SC in the hands of the same community that was to define SMS in the PLMN; i.e. to have defined the SMS-SC as being a part of PLMN.

The considerations above should not necessarily lead to the conclusion that DGMH made a suboptimal decision when it was free to pick the opposite. Regulatory regimes had already emerged that led the corresponding delegations to believe that implementation of SMS with an SMS-SC inside PLMN might be difficult or impossible. The above considerations therefore apply to the whole of the telecom community – including regulators.

4.6.4   What Interface to Choose for the SMS-SC↔SMS-GMSC/ SMS-IWMSC Connection?

As described above, the decision made for the SMS-SC↔SMS-GMSC/SMS-IWMSC connection was to offer an optional protocol based upon ITU's and ISO's ROSE given in an annex to GSM 03.40.

Retrospectively, the decision may not be judged to be a wise one. The protocol suggested in the annex to GSM 03.40 was later changed to one based upon Signalling System No. 7. However, its status remained 'optional', which has resulted in several alternative protocols developed by the vendors.

Obviously, the decision to locate the SMS-SC outside the PLMN cannot altogether be blamed for the decisions taken in the period from 1987 to 1990 concerning the SMS-SC↔SMS-GMSC/SMS-IWMSC connection. Even if the SMS-SC was located outside the PLMN, there was in theory nothing to prevent the bodies specifying GSM to define a mandatory interface based upon Signalling System No. 7 for the SMS-SC↔SMS-GMSC/SMS-IWMSC connection.

However, there are some aspects of locating the SMS-SC outside the PLMN that might be regarded as giving preference to the decision taken on the SMS-SC↔SMS-GMSC/SMS-IWMSC connection in 1989/1990:

- There was an unspoken policy to try to keep the access of Signalling System No. 7 inside the family of network operators. This policy has since been relaxed, but it was quite significant during the 1980s.
- The urge to reach quite tight deadlines for producing stable specifications for all of the functionality that contributed to the provision of GSM services made it tempting to suggest that interfaces to nodes outside PLMN could be subject to optional protocols.

If the SMS-SC had been defined as being part of the PLMN, it would probably have led to a quick decision in which a MAP-based protocol was defined as mandatory for the SMS-SC↔MSC connection. Routing might have been simplified, as much of the functionality of the SMS-GMSC and SMS-IWMSC related to SMS might have been included in the SMS-SC itself (interrogating the HLR, etc.).

4.6.5   Fixed Interworking or Variable Interworking?

During 1988 a question came up concerning the routing principles between the SMS-SC and the GSM PLMN. Did the routing principles have to be very restrictive, i.e. did the SMS-SC have to be connected to just one or to a small set of MSCs? Or could the SMS-SC be connected to a variety of MSCs in one or several PLMNs in such a way that the routing of a short message being transferred between the SMS-SC and a roaming MS could pick the MSC that would minimise the travelling distance? In DGMH terminology, the first alternative was denoted 'fixed interworking', and the second 'variable interworking'.

DGMH analysed this for approximately half a year. As already mentioned, the motivation behind variable interworking was to shorten the path that the short message had to traverse between the SMS-SC and the MS.

The conclusion was drawn at a DGMH meeting in Espoo, Finland, in February 1989. DGMH advised WP4 and GSM to stay with fixed interworking, mainly for the following reasons:

- As the SMS-SC will not be able to interrogate HLR and as in most cases there will be no intermediate network – ISDN or PSPDN – between the SMS-SC and the MSC, a possible scheme of variable interworking applies to mobile-originated SMS only.
- The benefit by obtaining a near-optimal routing in the mobile-originated SMS case was not found to outweigh:
  ✓ the increased complexity and cost related to variable interworking;
  ✓ the problems of combining fixed and variable interworking;
  ✓ the risk of barring or obstructing later enhancements of SMS.

The recommendation of DGMH was endorsed by WP4, and 03.40 was updated to cover fixed interworking only.

4.6.6   Some Concluding Remarks

The above four choices were some of the most significant that DGMH were confronted with during the first three years of its existence. If any experience is to be drawn from the choices we made and decisions we took, it must be the following:

- apply pragmatism – not only principles;
- take responsibility for service standardisation – e.g. by including all entities mainly relevant to the service in the scope of standardisation;
- give priority to simplicity at the expense of complexity.

To summarise, it was a wise decision to use the signalling capabilities of the GSM PLMN for short message transfer, thereby securing the possibility of a rapid and low-cost deployment of SMS after GSM network roll-out. It was definitely wise to restrict the standards to allow for fixed interworking only. However, it might have been a better choice to let the SMS-SC be a node within the GSM PLMN, implying that standardisation and specification of the SMS-SC would have been the responsibility of ETSI and later of 3GPP. That just might have enabled more powerful capabilities within the SMS-SC and thereby more powerful short message services, provided that the protocol between the SMS-SC and the MSC had been mandatory and unambiguously defined by the same standardisation bodies that were innovating the PLMN and the network nodes, i.e. ETSI and 3GPP. Finally, it must be concluded that leaving the interface between the SMS-SC and the MSC with an optional rather than a mandatory protocol was not a fortunate choice.

A vision of SMS that one might have expected to emerge already in those first three years of design was to expand the protocol of SMS-SC ↔ MS at the SM-TL by adding more PDUs and making use of compressed information elements in some of them. A straightforward example of means to make SMS a more powerful user-to-user communication tool is the short number option of the address field (see Section 4.5.7). The individual short message was not and is not defined to be able by itself to contain extensive addresses in the address fields (such as email addresses in the aa..a@bb..b format[17]) or extensive address structures (such as distribution lists). However, using, for example, certain PDUs for this purpose, both an email address and a distribution list may be established in the SMS-SC and connected to a short number, acting like an alias to the mobile user. In this way, mobile users might communicate to a variety of other users that they might not reach by regular addressing means – of course, still being subject to compromise solutions, e.g. in cases where the involved communication partners – the SMEs – in their responses might exceed the maximum number of characters or apply formats other than pure text.

---

[17] Email with the address format depicted above was passed between members of DGMH during its first three years, so the Internet form of email was definitely already recognised in Europe at that time.

4.7   Final Remarks on the Period of the First Three Years of DGMH

When looking back at a period of designing something new after a rudimentary requirement specification, one often tends to think: 'Why did it take so long?'.

This may also be the case with SMS by DGMH. In September 1990, after three years with more than 20 meetings, we had just reached a stage that might be characterised as a stable overall design. However, such a retrospective glance may often suffer from not considering every aspect of the outset and its surrounding conditions.

One of the major factors that may explain the time needed for the maturing of SMS is that, in contrast to most other GSM services, this was truly a new service. It had no other predecessor in PSTN, ISDN, CSPDN PSPDN or any other type of previous network. Along with deciding on every tiny little item of technical design, DGMH and WP4 had to discuss market aspects and make decisions on the basis of these. For most of the other services, the procedures were in many cases simpler: one just looked for what made the GSM service similar to the corresponding ISDN service.

It is fair to say that the design of SMS needed quite some time to mature, and that during the first three years of DGMH's lifetime this did happen. The number of WP4 members or other GSM experts who attended at least one of the 21 meetings of DGMH during those three years was close to 50 (see also Annex 4 for DGMH attendance). The number of people attending the first meetings was quite small, and there was no stable core attendance. From the summer of 1988 this changed. The interest in SMS in WP4 grew considerably, the perception of SMS as just the paging service of the GSM user gave way to a wider perspective and a more stable crew of SMS innovators emerged.

I would like to thank all of the almost 50 attendants for their contributions to the SMS design. In particular I would like to express special gratitude to the following people: Eija Altonen, Alan Cox, Kevin Holley and Didier Luizard. These four experts undertook responsibility for major work items that demanded great effort over a period of many meetings, and they always approached the challenges of SMS in a proactive way. Kevin, who succeeded me as both chairman of DGMH and editor of GSM 03.40, entered the DGMH at meeting 12 (Luleaa, July 1988) and stayed as a regular attendant for the rest of my period as chairman. His close attention to practically all issues on SMS that were dealt with by DGMH in that period provided for a more than sufficient continuity of the work that went on after 1990.

To sum up, the first three years wiped away the uncertainty of what the contours of SMS might be like. It vaporised the perception of SMS as consisting of two services – an essential mobile-terminated one and a not so essential mobile-originated one. From 1990 on, the perception of SMS was that it was a two-way service for which both directions were essential. The importance of this recognition can hardly be overestimated. From 1990 on, the basic characteristics of topology in terms of network, routing, protocols, protocol architecture and service elements were stable.

Still, there was to my knowledge no unrealistic assessments either in DGMH or in GSM4 that the 'job was done' and that the present material from the GSM groups could

be sent to vendors as fully completed and bug-free specifications to be used to establish
the production lines.

4.8   Work on SMS in GSM Bodies Outside GSM4

4.8.1   GSM 04.11

As GSM 03.40 only dealt with the protocol at the SM-TL layer and the protocol between
the SMS-SC and the MSC at the SM-RL, there was a need to specify, for example, the
transfer of the message at the SM-RL between the MSC and the MS (see Section 4.5.5).
Experts from the L3EG of WP3 took on that responsibility, and in the summer of 1988 the
first draft of GSM 04.11 – the recommendation of that protocol – was presented in meeting
12 in Luleaa, July 1988 (see Annex 3).

GSM 04.11 took substantial elements from GSM 04.08 owing to the similarities of setting
up a call connection for speech or circuit-switched data and setting up the signalling path
for short message transfer. GSM 04.11 was designed in a way that allowed for keeping the
signalling path until the transfer attempt was finished, so that either a confirmation or an
error indication might be returned without once again invoking the set-up procedure.

I would in particular like to thank the L3EG – in particular Jon Reidar Roernes and Knut
Erik Walter – for the efficient way in which they contributed to the swift establishment of
a first approved version of GSM 04.11.

4.8.2   Map Operations for the Support of SMS

After the discussion on what transport service to use for the transfer of short messages
between MSCs, and after the decision was taken to incorporate MAP operations to cater for
that, SPS-SIG was asked to elaborate a proposal for new MAP operations for that purpose.
SPS-SIG came back with an operation – *forwardShortMessage* – that might be invoked for
both transfer cases SMS-GMSC → MSC and MSC → SMS-IWMSC.

Together with the operations required for signalling purposes on the transfer of short
messages, the most important MAP operations related to SMS were:

- *forwardShortMessage*
- *sendRoutingInfoForShortMsg*
- *sendInfoForIncomingCall*
- *sendInfoForOutgoingCall*

See also Section 4.5.6.

I would like to thank WP3 and SPS-SIG – in particular Christian Vernhes, Bruno Chatras
and Jan Audestad – for the swift and efficient inclusion of the functionality required for
SMS in MAP.

4.9   Other Tasks of DGMH

The DGMH did not altogether deal with the short message service point-to-point, even if the work with that service gradually consumed most of the time of the drafting group. The DGMH was also instructed to design the services of MHS access and short message cell broadcast.

The GSM ambitions of MHS access in 1987 consisted of two services:

- basic MHS access;
- advanced MHS access.

Basic MHS Access was intended to provide a simple access mechanism to an appropriate point of the MHS system, which turned out to be the user agent.

Advanced MHS access was intended to provide a mechanism that would make the mobile user a more integrated part of the MHS system, and thereby make the access of MHS services more easy and less cumbersome than it would have been by means of basic MHS access.

The 1980s was the decade when the ITU and ISO made substantial efforts to establish a powerful and comprehensive message handling system by the X.400 set of recommendations. The first version of the standard to be commercially deployed was the so-called 1984 version. It had an amazingly large repertoire of services and service elements, but in terms of network nodes it consisted mainly of an MTA (Message Transfer Agent) and a UA (User Agent). The basic MHS access aimed to connect the GSM user to the MHS via the user's UA by access means such as PAD or X.32. The next significant upgrade of X.400, the so-called 1988 version, introduced another network node – the message store. Whereas the perception of the UA in the 1984 version was that it would reside in a fixed location and be connected to the MTA by means of high-capacity lines, the 1988 version opened up for UAs that might be mobile or nomadic and that would have to connect to the rest of the MHS via low-capacity data connections. The protocol between the UA and the message store – P7 – was designed on the basis of requirements that it could feasibly be deployed on a non-permanent and narrowband connection. This became the basic challenge facing advanced MHS access: would it be possible to align advanced MHS access with the concept of the UA $\leftarrow$ P7 $\rightarrow$ message store – where the UA resided in the mobile station – in the 1988 version of MHS?

DGMH and WP4/GSM4 gradually recognised two things:

- The scope of GSM recommendations and the scope of X.400 were quite different in terms of OSI protocol layers. Whereas the GSM recommendations dealt mainly with layers 1, 2 and 3, the X.400 recommendations dealt with layers 6 and 7. X.400 did not say much about the layers underneath, except that they should be based upon the OSI protocol stack, which for the network layer would mean X.25. For this and other reasons it became obvious that the most convenient way of accessing MHS for a GSM user would be simply to access a UA located outside the PLMN by means of the most feasible GSM data services as they evolved.

- The conviction held by the telecom community that X.400 was the message handling standard and service for the 1990s gradually faded. X.400 did not appeal very much to marketing people at either vendors or operators. It was huge and complex, and some of the user interfaces – e.g. the horrendous addressing formats – did not appear to be very user friendly.

Therefore, the DGMH proposed that the intended two GSM recommendations of MHS access be converted into a technical report describing how access to a MHS system might be achieved.

Cell broadcast was defined as a service by which a text message fed into the PLMN was to be transferred to all or a subset of the mobile stations temporarily under the radio coverage of all or a subset of the base stations of the PLMN.

Even though cell broadcast in the DGMH and IDEG/WP4/GSM4 never drew as much interest as the SMS point-to-point service, some supportive experts completed work to bring the service to full definition, which incorporated:

- the entry point at the PLMN for the text messages to be sent on a broadcast basis;
- addressing the set of mobile stations to receive the text message;
- the message format;
- the underlying protocol of the transfer over the radio interface (GSM 04.12).

In spite of the efforts in DGMH and GSM4, the cell broadcast service never became as popular as the SMS point-to-point. This lesson may be of interest for future service and product development, and one of the reasons may be as follows.

Business models based upon broadcast services like cell broadcast of GSM appear to be more difficult to make successful than those applicable to the point-to-point services. There are several reasons for that. First, the provider has to cater for some sort of selective acceptance of cell broadcast messages by those users that subscribes to the services. The complexity may be further increased by offering some dependency upon the location of the user, e.g. a user only accepting cell broadcast messages when being in certain regions. The content must be of such a nature that it is not critical that the user misses some of the cell broadcast messages that he should have received, e.g. due to being turned off, being out of coverage, etc. Finally, one has to wrap up the content within the framework of character strings of approximately the same length as the point-to-point messages. Thus, one should not be surprised that the cell broadcast has so far failed to penetrate the marketplace. However, the growing interest in providing public warning messages (e.g. terrorist attack, earthquake, tsunami warnings, etc.), where all mobile phones in a specific area can be targeted, may see a useful roll-out of cell broadcast.

5

The Evolution of SMS Features and Specifications from October 1990 to the End of 1996

K. Holley
Telefónica Europe

At the start of this period in the history of GSM, SMS was very much seen as a minor add-on, and much of the focus of the GSM community was on ensuring that voice worked well and on preparing for half-rate codec so that urgently needed additional capacity could be provided. Therefore, the focus of many of the reports of GSM is towards voice.

At the same time, the GSM4 community, focused on data services, considered SMS to be essentially complete for the first implementations, leaving only a few 'odds and ends' to tidy up for the basic design. The *baton* for chairing the DGMH (Drafting Group on Message Handling – the technical group responsible for SMS) was passed to me very suddenly. There were three or four of us in the room, and the GSM4 chairman, Graham Crisp, came into the room and told us that Finn Trosby would be unable to continue with his services as chairman. He said that we needed a new chairman for this work and explicitly cited Eija Thiger (Nokia) and Kevin Holley (Cellnet) as the experts with the most experience. Eija said that she did not want the job, and so there was only one candidate. Although by then I had been engaged in the standardisation activity for three years, this was my first leadership role, and there were still many discussions to be had. Often in standards we think we have completed everything necessary, only to find that new avenues open up and all of a sudden there is much work to be done!

Short Message Service (SMS)   Edited by Friedhelm Hillebrand
© 2010 John Wiley & Sons, Ltd

5.1   Topics Discussed in this Chapter

From 1990 to 1991:

5.2.1	Continuous Message Flow
5.2.2	Multiple Service Centre Scenarios
5.2.3	Delivery Reports
5.2.4	SMS Character Sets
5.2.5	SMS – an Optional Feature?
5.2.6	Storing SMS on the SIM Card
5.2.7	Unacknowledged SMS
5.2.8	Memory Capacity Available
5.2.9	SMS Negative Time Zone
5.2.10	Length of Binary SMS
5.2.11	Sending to and Receiving from Non-numeric Addresses
5.2.12	SMS API
5.2.13	Storage of SMS in the Phone
5.2.14	Upgrading Network Capability Reduces SMS Length

From 1992:

5.2.15	SMS to an External Terminal
5.2.16	Specifying Service Centre Interconnect to Cellular Network
5.2.17	Replace Short Message
5.2.18	Detecting Terminal Capabilities

From 1993:

5.2.19	Nokia Cellular Data Card
5.2.20	Improved Error Reporting
5.2.21	Receiving SMS from External Systems
5.2.22	Avoiding SMS Duplicates
5.2.23	Expanding SMS Character Capabilities
5.2.24	Using SMS to Alert to Waiting Voicemail
5.2.25	Which Time Zone to Use for SMS

From 1994:

5.2.26	Delaying SMS after Phone Power-On
5.2.27	Manual Flow Control
5.2.28	SIM Management by SMS
5.2.29	Icons for Voicemail Alert
5.2.30	Ensuring Proper Display Control via Control Characters
5.2.31	Concatenated SMS
5.2.32	SMS Divert

From 1995:

5.2.33	SMS Message Indication
5.2.34	More on SMS Alphabet Coding

5.2.35   Interworking with Email
5.2.36   SMS Between Networks
5.2.37   Use of AT Commands for SMS

From 1996:

5.2.38   SMS Compression
5.2.39   More on New Alphabets
5.2.40   More on International SMS Messaging

Abbreviations used in this chapter are explained in Annex 1. Sources for quoted GSM documents can be found in Annex 2, and a complete list of documents used can be found at www.GSM-history.org. An overview of the evolution of the key recommendation GSM 03.40 can be found in Annex 7.

5.2   Technical Improvements to SMS 1990–1996

The following improvements were made between 1990 and 1996. The text is developed from reviewing the meeting reports and input contributions from the time, along with personal memories of the meetings and in some cases a personal view of the impact these changes ultimately had.

5.2.1   Continuous Message Flow

My first meeting as chairman was in Paris in December 1990. There were four people attending this meeting full time, and another seven part time. Several issues were on the table. Significant attention was being paid to future traffic loads, and technical experts had identified that the transfer of many short messages to an individual mobile could be handled far more efficiently by sending all messages at once rather than clearing the radio channel after each short message transfer. This idea is a bit like the postman holding the letterbox open while he posts multiple letters through the letterbox, rather than opening the letterbox each time. In the case of SMS, opening the letterbox takes some time and prevents the use of radio resources by other users, so, in anticipation of future traffic loads, it was better to promote continuous message flow until all messages had been delivered.[1]

5.2.2   Multiple Service Centre Scenarios

Some aspects of the original SMS development had been made with the radio paging community in mind. A radiopager has a single server that sends all its messages, and it was not necessary to consider multiple servers. However, as SMS flows across national and international signalling links, it is possible, and indeed these days it frequently happens,

---

[1] GSM4 143/90.

that more than one source of SMS (more than one source operator) sends SMS to a given destination. The design alerts the source operator service centre when a phone is switched on. At first it was only possible to alert one source operator service centre. With international links and multiple centres needed per operator, it was clear that multiple alerts were needed, and these were included.

In addition, the inbound routing of a message might be different from the outbound routing. For example, a message sent from an Orange user to a Cellnet user would go via the Orange SMS Centre, but a reply would go via the Cellnet SMS Centre. But if the message originated from a third-party service (e.g. a sports news text service) connected to the Orange SMS Centre, then the Cellnet SMS Centre would probably not be able to deliver a reply. So the concept of 'reply path' was developed to indicate that replies should go back via the same route as the original message. The problem then was that the recipient would probably have no commercial relationship with the provider of the SMS centre (in the example, the Cellnet user with the Orange SMS Centre). So it was envisaged that the sending centre would keep track of the messages, relying on the reply path, and reverse-charge the reply message (Figure 5.1).[2]

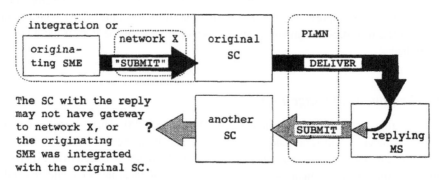

Figure 5.1   Multiple service centre scenario

In practice this feature has not been used in the mass market, and operators have collaborated to ensure that short codes work across every network in a given country.

5.2.3   Delivery Reports

Another enhancement to SMS was the ability to determine what happened to a message sent earlier. SMS was initially envisaged as being heavily used by the dispatch community, and the confirmation of receipt of a message was seen as important. But it was also seen as important to identify what had gone wrong with a message. The first deployment scenarios were expected to be like radio paging message centres, where an operator would type in a message dictated over the phone and submit that to the destination number. In order

---
[2] GSM4 184, 185, and 186/90.

that a message centre operator could advise later if the message was still in a queue for delivery because the mobile had been switched off, or had failed owing to timeout or to other circumstances, as well as indicate that the message was successfully delivered, some additional features were needed to capture and store this information.[3]

5.2.4   SMS Character Sets

The SMS character set was inherited from ERMES, which provides Greek letters in capitals; however, there were some incorrect characters in our tables, and we had to make some changes in accordance with the desires of the Greek operators.[4] These last-minute changes for the first implementation phase of GSM made common adoption of the character sets impossible for those characters, as some manufacturers had already implemented SMS and would not be able to change their initial handset design.

5.2.5   SMS – an Optional Feature?

Racal (to become Vodafone) pointed out that the SMS capability was defined as 'optional', which would be a serious problem.[5] It was envisaged that many manufacturers would omit to support SMS, which would make it hard for the market to take off, as no one would know a priori whether a given handset could receive the message. Even the introduction of a 'checking' mechanism seemed complex and expensive to deploy, and so our verdict was that it would be essential for all mobiles to be capable of receiving SMS from the earliest possible time. This push for all mobiles to provide support was at first rejected by other groups,[6] and it was suggested that the main creation of SMS messages would be by dictation to an operator over a premium-rate phone number and that the costs for this would be significant – if a subscription was required for SMS, then there would be no guarantee that the message could be delivered after the money had been spent creating the message. At the end of 1990, GSM4 sent a liaison including the text shown in Figure 5.2.[7]

```
It has been brought to GSM4's attention that due to the fact
that mobiles may or may not be equipped to receive short
messages, a shadow is cast over the commercial success of the
short message service.
```

Figure 5.2   Final clarification of the mandatory character of SMS

This liaison ultimately led to the clarification that indeed it was mandatory for manufacturers to implement mobile-terminated SMS, and later also mobile-originated SMS.

---

[3] GSM4 90, 198/90.
[4] GSM4 176/90.
[5] GSM4 162/90.
[6] SMG4 23/92.
[7] GSM4 199/90.

5.2.6 Storing SMS on the SIM Card

In the first release of GSM specifications, the message centre address for routing SMS messages was stored in the mobile device itself.[8] But this meant that a user would always have to configure the mobile device, and this could be complex. It would be far easier to have the operator preprogram the SIM card with the SMS message centre address, and it was agreed that this would be a useful enhancement in GSM phase 2.

5.2.7 Unacknowledged SMS

British Telecom made some proposals that a new version of SMS could be introduced, where the messages were not acknowledged.[9] A kind of 'cheap' version of SMS with no guarantees of delivery. The SMS community felt that introducing additional options like this added to the complexity of the system and would also add to user confusion if they could be assured delivery for some messages but not for others. In the end, this proposal was not adopted.

5.2.8 Memory Capacity Available

It may be hard to believe but initially the storage for SMS was very limited, with some handsets providing no storage at all and some early SIM cards only providing space for one message.[10] These were extremes, but often the end-user was limited to a low number of messages, in the region of 5–10. In such cases where the user was regularly receiving messages, it would be important to delete old ones, and yet there was no knowing how important a given message would be, so automatic deletion did not seem appropriate. So how could we ensure that the system could cope with limited memory at the mobile? Some early implementations coped with this by having a buffer of messages where the oldest read message would be deleted unless explicitly 'saved' (e.g. to the SIM card). But not every handset did this, and many would send back an error to the network saying that their memory was full. For the system to keep retrying such messages was the only option available at first, but this was highly inefficient if the user did not delete messages for a long period of time. So we decided to introduce a 'memory capacity available' indication. Once memory became available in the handset, a message would be sent telling the message centre(s) with waiting messages that they could retry. This was a more efficient way to handle the issue of limited memory.

5.2.9 SMS Negative Time Zone

It was not always possible to provide a neat way to structure the coding of SMS header information once the first implementations had been 'committed to silicon'. Unfortunately,

---

[8] GSM4 4/91.
[9] GSM4 7/91.
[10] GSM4 10/91.

this has led to some anomalies in the coding, perhaps typified with the Time Zone coding in the Service Centre Time Stamp. In 1991 it was discovered that the coding for Time Zone was ambiguous. The original intention was to provide a range to cater for all time zones in the world at a granularity of 15 minutes. But the specification could be read in such a way that negative time zones (west of GMT) were impossible to code. Luckily, phase 1 of GSM was restricted to Europe, so there was never a need for a negative time zone. This time zone coding caught out many manufacturers, and the first SMS service centres were transmitting incorrect time zones for many months, leading to confusion when a clever mobile interpreted the time zone according to the standard. The previously mentioned problems with the standard were solved in 1991.[11] Some mobile implementations ignore the time zone, and even today some mobiles use their own internal 'time of receipt' and do not show the 'time of sending'.

5.2.10   Length of Binary SMS

Another interesting coding issue was with the length of the user data field (the actual text of the message).[12] As 7-bit coding was used, the length was expressed in characters and not in bytes. This was fine for 7-bit messages, but in phase 2 the GSM community wanted to introduce binary messages too. It made no sense for binary messages to have a length in septets,[13] so we decided that the phase 2 specifications had to be explicit about whether the length was in septets or octets. There was obviously no header space to indicate this, so it had to be governed by the Data Coding Scheme information and we had to specify which Data Coding Schemes would have their length counted in septets and which in bytes. At this time there was a review of the length of SMS messages; this was included in the meeting report[14] and is reproduced in Figure 5.3.

5.2.11   Sending to and Receiving from Non-Numeric Addresses

Originally it was envisaged that SMS would be sent between terminals with just a telephone number, and this was the situation in GSM phase 1. But for phase 2 it was felt to be useful to be able to send to or receive from text-based addresses as well.[15] Again, we were limited by the requirement to keep 160 characters available, so we had to use the existing address space of 10 bytes. This effectively restricts the number of characters (encoded with 7 bits) to 11. But this still allows messages to be received from well-known names like

---

[11] GSM4 133/91.

[12] GSM4 126/91.

[13] Septets are sets of 7 bits – using 7-bit character coding allows the GSM standard to compress 160 characters (160 septets) into 140 bytes.

[14] GSM4 126/91.

[15] GSM4 58r1/91.

```
Annex 5 - Breakdown of SMS PDU lengths

Mobile Originated                           Mobile Terminated

User Data         140                       User Data         140
MTI,VPF,SRR         1                       MTI,MMS,SRI         1
Message Ref         1                       Orig Addr          12
Dest Addr          12                       Protocol Id         1
Protocol Id         1                       Data Code Schm      1
Data Code Schm      1                       SC Timestamp        7
Validity Per        7                       User Data Len       1
User Data Len       1

TPDU TOTAL        164                       TPDU TOTAL        163

TPDU              164
TPDU Length         1
RP-Message type     1
RP-Message ref      1
RP-OA,RP-DA        14

RPDU TOTAL        181
```

Figure 5.3 Breakdown of the SMS PDU lengths

Telefónica. There was even a discussion in 2009 on a podcast about the TV series 'Heroes'.[16] Here, the presenters debated whether someone could really reply to a text message from 'Unknown' without the receiver knowing the number. But, of course, that is possible with the alphanumeric encoding developed as a part of GSM phase 2 in 1991, with the address set to the text 'Unknown'. Support has to be enabled in the SMS message centre.

There was a debate at the time about the interaction between this feature and the Call Barring feature, which can be used to prevent SMS being sent to international numbers, for example. Also, there are issues with the Fixed Dialling Number capability on the SIM. But, in the end, these issues were not considered important enough to prevent the use of alphanumeric addressing in SMS. Today, this is used by some operators for sending out messages to subscribers, e.g. O2 (UK) subscribers receive messages when roaming abroad from O2Roaming.

5.2.12 SMS API

Application programming interfaces (APIs) are very important for machine-to-machine or telemetry applications as well as for machine-to-human applications. Originally it was envisaged that APIs would only be needed at the network side for SMS and that the messages would be displayed at the phone for consumption by the user, and entered at the keyboard by the user. However, after a while it emerged that there was a need for APIs at the phone side – to allow remote monitoring applications and control of SMS at the phone by external devices. The original idea was to extend the interface to a computer terminal by creating a protocol that would run over a serial cable and transfer messages to and from the

---

[16] 10th Wonder Podcast Live Show Replay 'Trust and Blood', around 18′10″ into the podcast.

phone. This was first discussed in a joint meeting between the technical SMS group and the requirements group in the summer of 1991.[17]

5.2.13   Storage of SMS in the Phone

Storage of SMS messages was originally envisaged to be in the SIM card; however, as time went on it became clear that the phones themselves had much more memory available and could store many more messages. Phone implementations started to allow messages to be stored on the phone. This was discussed between GSM WP1, the GSM requirements group and DGMH, the technical SMS group,[18] and it was agreed that, for reasons of privacy, it should not be possible to view the SMS messages stored in a phone after the SIM is changed. This feature was implemented by some manufacturers, but after a while they seemed to lose interest, and there are now many stories of private messages being found and published because the phones do not lock the messages to the SIM card as originally defined by GSM.

5.2.14   Upgrading Network Capability Reduces SMS Length

In December 1991 there was a challenge to the length of SMS messages as a consequence of progress in international signalling systems. The CCITT Blue Book overheads had led to the SMS length restriction of 160 characters, but the newer White Book version took more overhead, and it was reported that, after upgrading to White Book, the guaranteed length would only be 116 characters (Figure 5.4).[19]

> The maximum length of TPDU that MAP version 2 can support without segmentation is 126 octets (which equates to 102 octets for SMS characters or 116 characters; see attached calculation). The reduction in length has been caused by additions to the TCAP protocol for the white book version.

Figure 5.4   Reduction in the TPDU length of MAP to 126 octets in the White Book, 1991

Clearly, it would be unacceptable to the market to reduce the length to 116 characters in the name of 'progress', so some other solution had to be found. Fortunately, White Book signalling allowed segmentation of protocol transport units (signalling messages), which meant that more than one signalling message would be required in the network per SMS (only in the network and not over the radio). Interworking between a network supporting Blue Book and one supporting White Book would be complicated, but eventually this was resolved and SMS could continue with the 160 character length.

---

[17] GSM4 139/91.
[18] GSM4 139/91.
[19] GSM4 278/91.

5.2.15   SMS to an External Terminal

The original SMS service was seen as simply for display at the mobile phone, but with such a complex system there was scope for additional features and functions. At the start of 1992, a specification to allow control of SMS, including sending, receiving and interrogating the mobile, from an external terminal (e.g. a PC) was started (Figure 5.5).[20]

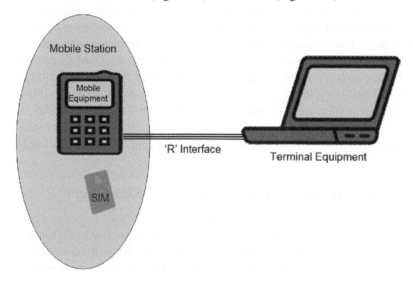

Figure 5.5   SMS to an external terminal

The design for this was based on an earlier contribution[21], which included the text shown in Figure 5.6.

The first version was based on the standard for control of modems in use at the time, called V.25bis, which enabled full control of SMS. This was later to be replaced by 'Hayes' AT commands. When considering what this would offer to the GSM system, it became clear that some SMS messages could be directed explicitly towards an external terminal instead of to the phone display. And, by extension, some SMS messages could be directed to the SIM card only. So changes were developed to enable this, including the ability of the mobile to alert the terminal to the existence of an incoming SMS message. The SIM specific message functionality paved the way for the SIM management capabilities used today (see Section 5.2.28).

5.2.16   Specifying Service Centre Interconnect to the Cellular Network

Meanwhile, a long discussion started on protocols for interconnecting SMS service centres with the cellular network.[22] The original X.25 specification was being supplemented by

---

[20] SMG4 22/92.
[21] GSM4 317/91.
[22] SMG4 64/92.

DGMH has discussed the use of the R interface to transfer short messages or cell broadcast messages from the MS to a terminal, or from the terminal to the MS (short messages only), and has come to the following conclusions:

- The control procedures should be defined as 'SMS-specific' procedures and should be similar to V.25bis.

- The default mode of the R interface should be extended V.25bis, with one additional command which changes the mode to 'SMS'.

- Transfer of messages should be achieved by taking SMS messages with header information as described in 03.40 or 03.41, adding an R-interface header and transferring this across the interface. The definition of this interface will be restricted to this transport, and will not take into account any character conversion from the GSM standard alphabet. This may be performed in the terminal, but this is up to the implementation.

- The protocol will be specified such that messages existing in the MS and/or new messages received over the air interface may be transferred to the terminal. Messages received from the terminal may be stored in the MS memory or directly transmitted to the Service Centre.

- An SMS protocol identifier will be specified which indicates that the message is destined for the terminal only, and not the mobile. The 'R' interface protocol will also permit existing messages in the MS to be transferred or copied to the terminal.

- A new alphabet identifier will be reserved for 8 bit transparent data transmission (SMS). A new language identifier / language group will be reserved for the same purpose (CBS).

Figure 5.6   Principles for the interface to external equipment

other proposals, including the use of MAP. The process being followed was largely the incorporation of a specification developed outside ETSI into 03.40, and 03.40 was becoming larger and larger. After debate at several meetings, it was eventually decided not to specify any of these examples in 03.40 and instead to write some simple high-level statements about requirements on these protocols.

5.2.17   Replace Short Message

Some more interesting capabilities were introduced in 1992, including the Replace Short Message function. This allows a message to replace (overwrite) one that was sent earlier.[23] There are seven types of Replace Short Message – a message tagged with one of these types can only replace a message with the same type. A message is only replaced if the original source address matches the one associated with the replacement message. And the strangely named Replace Short Message Type 0 was also defined – this message is silently discarded at the mobile (although acknowledged). Type 0 messages can be used to determine that the mobile is switched on and in coverage and, for example, available for an incoming call. As SMS has the Alert function, a system can be advised the moment a mobile becomes

---

[23] SMG4 116/92.

available. This has been used in services such as Callback Voicemail – dialling a mobile to
play waiting voice messages when it is switched on.

5.2.18   Detecting Terminal Capabilities

There was a proposal from Detecon to make a mobile detect and reject features to be
defined in future versions of 03.40, to allow a sender to know that the recipient was
unable to process the message properly.[24] Unfortunately, this would result in an automatic
(charged) SMS response to the incoming SMS, and for this reason the proposal was not
adopted. However, this feature would have been useful, for example, in determining that
the recipient could not read an incoming SMS message because, for instance, it used a new
encoding method for the message characters (see later about the introduction of Turkish
characters).

5.2.19   Nokia Cellular Data Card

The work on the specification for external terminals had started in 1992, but in 1993 there
were several additions and corrections from Nokia.[25] Looking back, it was clear that they
had started to implement the protocol and were fixing problems they had found in implementation. The Nokia Cellular Data Card was the first implementation of the standard SMS
control protocol in a PCMCIA card.

5.2.20   Improved Error Reporting

During 1993 there were other additions made to SMS to provide a smoother user experience.[26] One example was a review of the possible error situations. It was agreed that
the error situations were not sufficiently handled in the standards and that therefore some
additions were needed so that the user could be informed of the type of error that had
occurred (congestion, problem with the destination number, etc.). Rather than calling for a
second message to be sent to the mobile with the SMS send result, the result was integrated
with the response coming back from the SMS service centre. The problem was that this
meant adding extra data in the lower layers defined outside SMG4. So SMG4 had to liaise
with the other SMG groups, and especially with those working on the network-to-network
protocols (MAP) and radio interface protocols, to ensure that this would all work properly.
The external terminal interface also had to be updated to take this into account. Finally,
there is a rich range of error codes, but these are unfortunately rarely implemented, with
the mobile phones just telling the user something like 'message failed'.

---

[24] SMG4 157/92.
[25] Various SMG4 documents from 1993, including 34, 69 and 185.
[26] SMG4 187/93.

5.2.21   Receiving SMS from External Systems

In 1993 the implementation of SMS was well under way, and many manufacturers and operators were interested in being able to connect a fixed network system to the SMS service centre for the sending and receiving of SMS.[27] In today's world this would have been straightforward and probably based on XML, but in 1993 there were few precedents for business-to-business data exchange, and those that were around were based on the packet-switched X.25 protocols. SMS was relatively new and had many features not seen before, so a richer protocol was needed. As many manufacturers had already specified their own protocol when the issue of standardisation was brought up, it was rather difficult to decide on only one protocol, and creating another new protocol would have taken a long time and introduced yet another option. So it was decided to create a technical report that listed all the various protocols that were provided by manufacturers, and incorporated the details of them all.

5.2.22   Avoiding SMS Duplicates

As a result of a flurry of SMS testing during 1993, it was discovered that retry attempts for SMS messages were resulting in duplicates.[28] This was felt to be a significant problem, but it was recognised that it would be difficult to detect a duplicate in the mobile-to-land direction. In the land-to-mobile direction there is a time stamp as well as a message reference, and this is sufficiently unique to detect duplicates (except where the first copy of the message is deleted from the mobile, and then the mobile cannot detect the second copy as being a duplicate of the first copy). In the mobile-to-land direction, the SMS service centre assigns the time stamp and the message reference is created by the mobile. It was discovered that, for retries of the same message, in some implementations the mobile was creating a new message reference, which meant that the duplicate could not be detected at the service centre. This was fixed by some tightening up of the wording in the standard, but also by discussion in the committee and letting all the manufacturers know the right behaviour for the mobile.

5.2.23   Expanding SMS Character Capabilities

In 1993 the spread of GSM was becoming significant and the SMS service needed to cater for many more countries than the original European ones. The first input asking for additional character sets came from Modarabtel in Tunisia.[29] The original SMS character set provided the main European characters as it was based on the European Radio Paging set. So there was no way to display any Arabic characters. The standardisation of additional characters took a very long time, largely because the participants in the meeting were not

---

[27] SMG4 155/93.
[28] SMG4 256/93.
[29] SMG4 212, 218/93.

able to discuss and agree what was needed in the non-European countries, as they did not have the expertise, and there was no participation from the countries that needed additional characters. One particular problem that was pointed out was that some languages write text from right to left, and some from left to right, and the interpretation of direction is very important when taking telephone numbers from SMS and using them to make calls. Without the required expertise in other languages, it would be difficult to ensure that the new languages were introduced in a harmonious and inclusive way.

Several solutions were proposed to the missing character problem:

- Introduce additional code page tables – this was felt to be rather onerous on the mobile manufacturers. Memory was very expensive, and to include lots of code page tables would simply cost too much. There was a tentative mechanism proposed for switching between tables, but this was abandoned when it was clear that there was a memory problem with lots of tables. It was also suggested that changing the language of the handset could cause the code page for SMS to change, but this would mean that people who speak more than one language and can receive SMS in a different language from the one set for the handset would receive strange-looking messages.
- Use the already existing mechanism for adding accents to characters, based on the T.51/T.61/T/52 standards – this was felt to be too character hungry, as any language where a significant number of characters have accents (e.g. Hungarian) would be using two SMS characters per displayed character, or in some cases three SMS characters per displayed character, making a big dent in the original 160 character length.
- Introduce a 14- or 16-bit character set – which could cover most of the characters world-wide.

As it was envisaged that the work of introducing additional character sets would be an ongoing exercise, it was agreed to move the character set specification material from several places in the main body, at first to an annex of 03.40, but eventually it was agreed to move this material to a new specification 03.38.

5.2.24   Using SMS to Alert to Waiting Voicemail

Siemens introduced the idea of a special SMS message to indicate waiting voicemail.[30] This was based on the Replace Short Message idea, which had been introduced in 1992. The initial reaction from the group was that SMS already had the ability for a service centre to send a Replace Short Message, and the text could indicate that there was a waiting voicemail. But after more discussion it was realised that special handling was needed to differentiate this type of short message from other types within the handset and to prioritise such messages in the view to the user. As this was for an indication about waiting voicemail, in effect a prompt to the user to dial in to collect voicemail, it was decided that Replace Short

---

[30] SMG4 223/93.

Message Type 8 was the wrong name for the code point. Other names such as Callback or Recall were dismissed because they were names of voicemail services provided by Cellnet and Vodafone respectively. It was agreed to call the code point Return Call Message.

5.2.25 Which Time Zone to Use for SMS

The Time Zone coding started to present a problem also in 1993 as a result of implementations and testing.[31] Manufacturers were unsure whether the time was supposed to be GMT or SMS service centre local time (it was supposed to be local time). In addition, the text referred to 'two's complement' coding and using a bit for the sign of the time zone, which were not compatible. So the text was clarified. In addition, the original limit of GMT $\pm$ 12 hours was lifted to allow for times outside this (e.g. New Zealand is on GMT + 13 with Daylight Saving Time). Many manufacturers struggled with the correct implementation of the time and time zone to start with, but the mainstream manufacturers soon corrected any faults. Today, many handsets display the correct time and adjust for the local time zone set. But some ignore the time stamp and even display the time the message was received, which is not very helpful if the phone was off when the message was sent and the message says something like 'meet you in an hour'!

5.2.26 Delaying SMS after Phone Power-On

1994 was another year where implementation issues were brought to the standardisation committee.[32] When a GSM phone completes a session on a radio channel, it is required to perform a rescan of the radio and re-read the broadcast information (which tells the phone about the network, the features available, the requirements on the phone, etc.). This takes a finite time (estimated at around 2 seconds). If during this time the phone is paged, then the page can be missed. So implementations with SMS were finding that, when a phone was switched on, it sent a location update message into the network. This triggered the sending of waiting SMS messages, but when the SMS page arrived the phone was busy rescanning the cell. While this was a significant issue, it was felt that the solution lay in introducing a timer in the SMS service centre that could be adjusted. The timer would delay the sending of waiting messages until 1 or 2 seconds after the alert had arrived at the service centre.

5.2.27 Manual Flow Control

A proposal for manual flow control (stop sending, restart sending, send me all pending SMS)[33] was not agreed in 1994 because there was insufficient support from the SMS community in SMG4.

---

[31] SMG4 245/93.
[32] SMG4 96/94.
[33] SMG4 120/94.

5.2.28   SIM Management by SMS

A proposal to use SMS for the purposes of SIM management was agreed to be beneficial.[34] There was some concern about earlier mobiles receiving such messages and storing them, available for the hacker community to modify and reuse, so further study was requested of the security and service design experts.

5.2.29   Icons for Voicemail Alert

A liaison statement received from the services group suggested that a special icon be introduced on the display to indicate that voice messages are waiting.[35] Although in principle the existing Return Call message coding could be used for this, the connection of the received SMS to the icon would be proprietary, so it was felt that this would be a useful change. SMG4 decided to introduce such an indication, called voicemail waiting indication, but also to include some text in the SMS message for backwards compatibility so that, if the handset did not support the indication, then the text would tell the user what to do.

5.2.30   Ensuring Proper Display Control via Control Characters

The special control characters CR (Carriage Return) and LF (Line Feed) have been in use since printing teletype terminals were used in the 1970s. CR has always meant to return the print head to the first character of the current line (allowing overprinting), and LF has always meant to move the print head down one line to the current position on the next line. However, in the mobile case, some manufacturers had interpreted these control codes differently.[36] To stop the text on the current line and restart at the first position on a new line, it is necessary to send both CR and LF. To ensure that the mobile manufacturers appreciated all of this, some minor changes were made to the wording.

5.2.31   Concatenated SMS

In a messaging system such as SMS there are times when a single message cannot convey everything needed. For such occasions it was felt useful to be able to string together two or more SMS messages in a 'concatenated' fashion. Some non-standard methods had already been in use for this, such as ending a message with the + sign and beginning the second message with the + sign; however, a standardised method that could be seamlessly processed by the sending and receiving mobile was attractive.[37] The header space in the SMS message was rather limited, and introducing coding for concatenated SMS was impossible without

---

[34] SMG4 122/94.
[35] SMG4 109/94.
[36] SMG4 194/94.
[37] SMG4 189/94.

shortening the SMS text. In addition, backwards compatibility was viewed as essential. So it was decided to introduce a clever scheme developed in the committee where the SMS header would point to part of the text of the SMS message as being part of the header information ('user data header'). An older mobile would see some strange characters at the beginning of the SMS message, but a later mobile would skip over these characters and display the real message. To make life as easy as possible for the poor user with an older mobile, the scheme ended the binary part of the message with a CR character. This meant that, if the user data header was short enough (fewer characters than one line of text), then the mobile (according to the standard definition for CR) would go back to the beginning of the current line and overwrite the binary characters with the real text, making the header invisible. For concatenated SMS the scheme was introduced where up to 255 messages of 152 characters (eight characters used for the header information) could be concatenated. This scheme has been widely implemented, at least up to six or so messages, and, although sending more than 160 characters will cost the price of two SMS, this is a useful feature and is much used in the market for longer messages.

5.2.32   SMS Divert

Diverting short messages from one mobile to another has been a topic of interest since 1994. The first attempts at creating a divert capability were with the use of SMS commands – special SMS messages sent to a service centre for the service centre to interpret. However, the pre-existing divert commands for other services used supplementary service *# codes and were sent to the Home Location Register or Visitor Location Register. If SMS commands were used for divert, then only the SMS service centre that received the SMS command would action the divert, meaning that SMS messages sent from other places would go to the original handset. Imagine T-Mobile subscribers who can divert their messages to another phone for messages received from T-Mobile customers, but not for messages received from Vodafone customers. Also, what should happen with Type 0 SMS messages (described in Section 5.2.17 above) which should not be diverted? Study was also suggested for Call Diversion Override and Loop Avoidance. Further work was clearly needed, but in the end this problem proved too complex and costly to solve for the perceived benefit.

5.2.33   SMS Message Indication

Earlier, in 1994, we had introduced the capability of Return Call Message which allows an indication of waiting voicemail. In 1995 we received a request from the services group[38] asking to be able to provide an SMS message indication, which could be used to place icons on the display for waiting voice, fax or email messages. A solution to this[39] was

---

[38] SMG4 9/95.
[39] SMG4 70/95.

found by adding User Data Header fields to include the type of message waiting and the number of messages waiting. This would then allow innovative solutions to be developed by the mobile manufacturers. This was then left for one meeting. At the following meeting, concern was raised by One2One[40] that strange characters would appear on the front of the messages with some mobiles, and instead a solution was proposed that used the originating address to switch on or off the icons. However, this solution was not found to be agreeable because the originating address was not intended for this purpose and because third-party voicemail operators had to be accommodated. The meeting agreed on a set of requirements (Figure 5.7) and for further offline discussions.[41]

It was agreed that the following requirements should be taken into consideration by any new proposal:

* Freedom to use the text of the message or not
* A mechanism to set an indication
* A mechanism to clear an indication
* Freedom to use the Originating Address as required for backward compatibility
* SIM storage kept to a minimum
     Replace Short Message scheme should work
     Newer mobiles should store nothing
* A field is required to say "store the message in the SIM anyway"
* A field is required to indicate voice, fax, email and other types of indication
* Backwards compatibility is very important
* Manufacturer option to allow easy calling of the number in the Originating Address field

The following people expressed an interest in fax discussions between now and the next meeting Messrs Gidlow, Trouwee, Meuronen, Barnes, Harris, Erskine and Holley.

Figure 5.7   Requirements for SMS message indication

The discussion continued at later meetings, with the introduction[42] of the use of Data Coding Scheme code points to indicate switching on or off of icons.

5.2.34   More on SMS Alphabet Coding

Earlier discussions about the introduction of new language capabilities for SMS continued in 1995. T-Mobile was an advocate of the standard T.51 mechanism to switch between language tables,[43] while input from Hungary[44] suggested more to introduce a new GSM-specific code table with an interesting scheme for 'locking' or 'non-locking' shift into this new table for specific Eastern European characters. A final decision was not made at this meeting because it was felt that further discussion was needed.

---

[40] SMG4 91/95.
[41] SMG4 Plenary Meeting Report, Uppsala, Sweden, 30 May–2 June 1995.
[42] SMG4 235/95.
[43] SMG4 3/95.
[44] SMG4 55/95.

During the period from 1992 to 1996, GSM was spreading to countries with non-Latin alphabets like the Arabic states, Iran and China. In China, GSM was competing with CDMA, and very soon there were hundred of thousands of Chinese users. The then SMG chair, Philippe Dupuis, therefore thought it was urgent to offer them the possibility of exchanging SMS in their own language. At that time Mr Dupuis was taking part in the promotion of GSM by speaking at various conferences in Singapore, Hong Kong and Beijing. At one of these conferences, possibly in Singapore early in 1995, a Chinese engineer from Motorola in Beijing came to him at a coffee break and explained that the use of Unicode would be an excellent solution. Mr Dupuis had never heard of Unicode, as there was very little exchange between telecommunications and information technology experts.

Back home Mr Dupuis made a quick enquiry, obtained a copy of the UCS2 alphabet and became convinced that it was something serious. At the next SMG meeting we adopted the specification enabling the use of UCS2. A few weeks after, Mr Dupuis included the information in his report to a GSM MoU meeting, probably in Rome. After this presentation, three participants from China, Taiwan and Hong Kong came together to Mr Dupuis and explained that they were very happy with this.

A few weeks later, in Beijing, Nokia demonstrated the transmission of SMS in Chinese ideograms.

During subsequent DGMH meetings there were further proposals relating to T.51 and GSM-specific code points, but in June 1995 a proposal from Motorola[45] noted that many computer systems were moving to adopt Unicode – a 16-bit coding system that included code points for most of the characters in use worldwide. This would result in the length of SMS being only 70 characters, but 70 Chinese characters would convey quite a bit of information, and for those languages with more characters per sentence a compression scheme would be able to improve on the number of characters per message. This proposal was adopted in 1996,[46] and the topic of international character sets for SMS became quiet for a few years (See Chapter 6 for later developments). The relevant SMG report[47] shows the adoption (Figure 5.8).

Additional alphabets for the Short Message Service

SMG approved the specifications relating to the support of UNICODE/UCS-2, the multilingual character set developed by ISO/IEC. This will enable short messages to be sent in characters of all world languages. The specification work is completed, but some consideration must now be given to compatibility issues for old mobile stations.

Figure 5.8   Additional alphabets for the SMS

[45] SMG4 145/95.
[46] SMG P-96-073.
[47] SMG RP_S_17.

5.2.35   Interworking with Email

Although X.25 message handling systems had been considered as a part of GSM, by 1995 the Internet email systems were starting to become popular and the interworking between these and SMS needed to be included. A paper from BT[48] proposed this (Figure 5.9).

1.   Internet mail interchange

The PID in the header can cope with ERMES pagers, X.400 mail and "message handling" centres, but not the most widely used electronic mail - Internet mail. This should be included. It is therefore proposed to use reserved codepoint 001 10010 to indicate translation to and from Internet mail.

Figure 5.9   Internet mail exchange

This was agreed as useful and incorporated. Subsequently, in 1996,[49] more descriptive text was added to specify how the subject of an email message could be included in an SMS message, enclosed by brackets or by # symbols.

5.2.36   SMS Between Networks

SMS was becoming more and more popular, but the users did not want to be restricted to sending SMS only within the network to which they were subscribed. A set of ad hoc interconnect possibilities had been formed accidentally because roaming agreements allowed the passage of signalling from one network to another, and this meant that SMS delivery was possible between two networks with a roaming agreement. Corridor discussions about defining a 'proper' SMS interconnect mechanism were taking place, but BT provided the first paper input[50] identifying possible mechanisms, concluding that the best mechanism would be the one already provided, and allowing signalling interconnect between all networks (Figure 5.10).

5.2.37   Use of AT Commands for SMS

During the March 1995 meeting in Düsseldorf, Germany, it became clear that some manufacturers had been developing a set of AT commands for GSM in competition with the standard V.25bis commands already included in the GSM specification. As always with standards, it is better to admit when an existing document has become less useful because the industry wants to use something else. In this case the industry really had moved away from V.25bis control and wanted to use AT commands. Also, for SMS the manufacturers were reluctant to continue the SMS Block Mode, despite the existing implementations (e.g.

---

[48] SMG4 115/95.
[49] SMG4 544/96.
[50] SMG4 237/95.

There are really only three ways to provide "internetwork" SMS:

1. HLR interconnect so that any SMS-GMSC can send to any subscriber.
2. SC interconnect via an undefined, non-standard interface
3. Submission to remote user's SC (origination to any SC).

(3) is very difficult to use, as customers would have to keep changing the SC address in your phone and/or SIM, *and* know all the SC addresses of all their correspondents as well as their phone numbers. You don't need to know that for voice calls, so you should not have to put up with anything else for SMS!

(2) is a nightmare for the network operator because it has to programme all the routing information for all SCs worldwide. It also uses a yet-to-be-developed interface.

(1) uses completely standard interfaces, was foreseen in the standards (i.e. that any subscriber can receive messages from any SC) and is extremely flexible. This is what should be implemented.

Figure 5.10   SMS between networks

Nokia Cellular Datacard). Instead of allowing the industry specification for AT commands to continue outside the standards, this specification material was brought into SMG4. At the time, the discussion shown in Figure 5.11 took place in SMG plenary.[51]

The liaison to SMG on GSM TS 07.05 and 07.06 was discussed. Several papers were dealt with at the same time (*Tdoc SMG 234/95 and 337/95*) as they are tackling the subject. Due to availability on the market of modified Hayes AT commands developed by Nokia, Ericsson and Hewlett Packard together and proposed that to SMG4. It was explained that this was filling some gaps in the standards. It was agreed to request SMG4 to introduce this in the standard taking into account the available documents from the proponents and specifically study the optionally of the commands. The question on whether or not the interface between the Card and the ME shall be standardised was raised. As it was impossible in the past to standardise this interface it was decided not to do anything due to the impact that this might have on the cost.

Figure 5.11   Use of AT commands

For SMS the AT commands were included in 07.05, which had originally only included the SMS Block Mode. Two further modes of operation were defined: the Text Mode and the PDU Mode. The Text Mode allowed AT commands to send characters of the SMS message as text, whereas the PDU Mode allowed more control over the header fields and did not require translation facilities between characters and an SMS message in the mobile. Thus, the PDU Mode was more suited to a relatively passive datacard implementation with the intelligence for creating SMS messages in software in a PC. The Turin, Italy, meeting of SMG4 in September 1995 agreed the first consolidated version.[52]

---

[51] SMG RP_S_14.
[52] SMG4 238/95.

5.2.38   SMS Compression

The discussion about the use of Unicode to represent characters, and also market input[53] about applications needing just a little more than 160 characters, led to the detailed investigation of compression mechanisms. Compression of 8-bit data was not found to work so well to compress 7-bit characters, so there was a second proposal[54] for a different scheme to compress the GSM default alphabet. The use of language-specific dictionaries would improve the compression ability further, as the graph from this second proposal indicates (Figure 5.12).

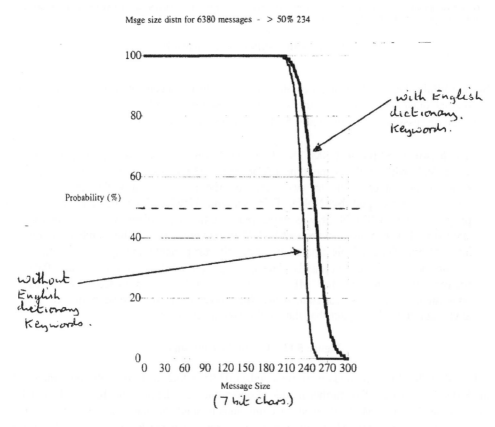

Figure 5.12   Efficiency of SMS compression

The technique to be used was further described in another document;[55] however, the feature was not finally complete until after the 1990–1996 period (see Chapter 6 for further information on SMS compression).

---

[53] SMG4 166/95.
[54] SMG4 353/96.
[55] SMG4 507/96.

5.2.39   More on New Alphabets

Motorola provided more information[56] in 1996 about the use of Huffman coding to compress SMS messages and identified that up to 50% compression could be achieved. They also provided a computer program for individual companies to test their own text with the compression algorithm. In addition, it was suggested[57] that mobile capabilities could be negotiated between SMS service centre and mobile handset, but this latter idea proved to be too difficult to develop because of the store-and-forward nature of the messages, and the need to modify the GSM Mobile Application Part for international signalling links.

5.2.40   More on International SMS Messaging

France Telecom started a new discussion in 1996,[58] indicating that there are security problems with the contemporary SMS interconnect possibilities and suggesting that all SMS should be sent via the 'home' SMS service centre. This would have caused a significant redesign of SMS functionality to provide service centre-to-service centre interworking and been expensive to deliver. In addition, there were other possibilities such as introducing an SMS policing function in the home network to which SMS messages are diverted. The cost of this would be met by the operator desiring a higher security, and the originating operator would not have to make any changes. In the end, the France Telecom proposal was not agreed. It is interesting to note that, when MMS was developed in 1999–2002, interconnect between operators used a service centre-to-service centre approach rather than the direct messaging capabilities of SMS. While this provided a more secure environment, it meant that MMS interconnect was incredibly slow to take off because it did not work until a specific agreement was drawn up and signed. SMS interconnect worked without any specific agreement, and therefore it was much faster to market between operators who already had roaming agreements.

5.3   Concluding Remarks on the SMS Period 1990–1996

At the start of this period the SMS standard was more or less fully defined from a theoretical perspective. Yet it was not really 'market ready' because we were not quite sure how the market would respond to the features available, and how much more development would be needed to meet the perceived market needs. Re-reading through all the old documents made me realise just how naïve we were about what would be useful. But at the same time we introduced features that are key market requirements today, such as SMS delivery reports. In development and testing there were many issues that came up, such as ensuring that SMS is delivered as early as possible without duplication. At the end of this period, SMS emerged as a strong and essential component of GSM, waiting in the wings for the text explosion to happen later in its life.

---

[56] SMG4 34/96.
[57] SMG4 36/96.
[58] SMG4 87/96.

6

The Evolution of SMS Features and Specifications from the Beginning of 1997 to Mid-2009

I. Harris
Research In Motion

Between 1997 and towards the end of 1998, SMG4 continued its work on SMS enhancements that had begun during the previous period described in Chapter 5. The most significant enhancements were 'SMS to email interworking', 'SIM toolkit' and 'SMS compression'.

Under ETSI TC SMG, SMG4 maintained its responsibility for SMS standards until reorganisation to 3GPP took place in 1999. Under 3GPP, a 'terminals-specific' group was established called TSG T, which comprised three technical subgroups T1, T2 and T3.

T2 was assigned the responsibility for continuing much of the work previously done by SMG4, which included SMS.

DGMH was renamed SWG3, a subworking group of T2.

The chairmen of SMG4 (and its SMS subgroup DGMH) and the chairmen of T2 (and its SMS subgroup SWG3) for this period are shown in Chapter 3.

In the transition from SMG4 to 3GPP T2, the main SMS specification ETSI TS 03.40 was given a new document number 3GPP TS 23.040. Similarly, ETSI TS 03.38, which defined the 'SMS Data Coding Scheme and default alphabet', was renumbered 3GPP TS 23.038.

Throughout the period between 1997 and 2009, SMS was continually enhanced, much of the work being based on the commercial need for improvements and the addition of features. Although approximately 26 new features were introduced during this period, the following ones, in chronological order, are considered by the author of this chapter to be

Short Message Service (SMS) Edited by Friedhelm Hillebrand
© 2010 John Wiley & Sons, Ltd

the most significant in terms of their value to SMS:

- SIM toolkit data download and secure messaging, 1997;
- SMS compression, 1997;
- Enhanced Messaging Service, 2001;
- Voicemail management, 2004;
- Routers, 2007;
- Language tables, 2008.

6.1   SIM Toolkit Data Download and Secure Messaging[1]

During the late 1990s, the need for a mobile network operator to remotely manage the SIM card began to emerge. The SIM card is the network operator's means of administering mobile phones on their network.

The complexity, frequency and amount of data required to be managed or updated on the SIM was small, and so SMS was identified as being a cost and operationally effective means of providing SIM management.

The use of this feature is transparent to mobile phone users and in no way affects how they use SMS. However, its importance in terms of the use of SMS does have an impact on mobile phone users by greatly reducing the need for making specific arrangements for their SIM to be managed that could necessitate frequent fitting of revised SIMs or visits to their service provider.

Remote SIM management is in wide use today. Two such applications are SIM data download and SIM toolkit.

6.2   SMS Compression[2]

The need for SMS compression had already been discussed in SMG4 in 1996.

The growth of commercial services during the late 1990s (which is explained more fully in Chapter 7) gave rise to concerns about the limitations of 140 octets (= 160 characters coded by 7 bits per character) for a single short message.

Generally speaking, every single short message has a cost to the consumer, and so, for commercial applications where potentially large numbers of short messages may be required, the need to keep the number of messages to a minimum was paramount. See also the section on billing in Chapter 7 (Section 7.10).

A mechanism for short message concatenation had already been defined in the standards prior to 1997, which provided a technical solution for those applications and uses where messages could not be easily conveyed within 140 octets. However, the cost of sending

---

[1] Introduced with Change Requests to TS 03.40 in SMG documents P-97-696 and P-97–918, and with a Change Request to TS 03.38 in SMG document P-98-096.
[2] Introduced by a new specification TS 03.42, a Change Request to TS 03.40 in SMG document P-97-058 and a Change Request to TS 03.38 in SMG document P-99-061.

multiple segments of a concatenated message is much the same as the cost of sending multiple SMS messages.

SMS had long provided the capability of binary encoding short messages. Such encoding is capable of containing messages with a high data content (particularly complex numeric messages) within a single short message.

There were however, a number of applications where neither concatenation nor binary encoding was a viable solution, and so attention was turned to 'compression'.

In 1996, Vodafone developed a means of compressing short messages, and TS 03.42 was created. Test vectors were also supplied to allow implementers to test their compatibility with other implementations. For those familiar with the history of ETSI document numbers, they will see that the number 03.42 was originally assigned to the technical report 'Advanced Data Message Handling System' (MHS) TR 03.42. ETSI took the unusual step of using the same number for the document 'SMS Compression' (TS 03.42).

TS 03.40 and TS 03.38 were revised to include the provision for compression.

The compression algorithm, based on a Huffman code, takes into account character frequency occurrence, character groups and punctuation. Typically, it is possible to compress 240 characters into the 160 character space available. The compression mechanism is also applicable to concatenated short messages, UCS2 encoded messages and binary encoded messages.

For most European languages, the compression efficiency is approximately the same for each language, but an additional option to include 'keywords' is defined for each specific language, thereby further increasing the compression ratio.

Compression is an option that requires specific software in the sending and receiving mobile, and so its use is largely limited to commercial applications where there is specific control over the receiving and sending devices. It is rarely used for casual consumer-to-consumer general text messaging.

TS 03.42 was renumbered to 3GPP TS 23.042 in the transition to 3GPP.

6.3   Enhanced Messaging Service (EMS)[3]

During the early 2000s, work had already begun on specifying a Multimedia Messaging Service (MMS) – a media-rich messaging service that was viewed by some as moving the need for messaging into the twenty-first century and likely to bring about the demise of SMS.

The complexity, implementation costs and consumer costs for MMS were predicted to be significantly higher than for SMS, but so confident were the protagonists of MMS that provision was made to encompass SMS messages within MMS to ensure smooth transition to the media-rich service.

However, there were also a large number of organisations who were not so sure that MMS was about to replace SMS but who also realised that SMS lacked lustre, particularly for young users who had stimulated an unprecedented growth of SMS. Nokia had attempted

---

[3] Introduced with Change Requests to TS 23.040 in 3GPP documents TP-000024, TP-010128, TP-010149, TP-020015 and TP-020079.

to make SMS more compelling by developing 'picture messaging' which allowed simple graphic images to accompany short message text. This was a proprietary solution available for Nokia mobile phones only.

In 1999, a Korean company called Bijitec came to 3GPP with a proposal to allow the user data field of SMS to include coding that would allow simple images, simple animations and sounds to be included with text or sent instead of text.

Clearly, with only 140 octets of user data available per single short message, the need to efficiently encode images, animations and sounds was paramount in order to keep the number of short messages and hence consumer costs to a minimum.

Other companies, Nokia, Ericsson and Motorola, came forward with alternative encoding schemes to challenge Bijitec. All four companies were given the common task of encoding a range of images such as a bee, a flower, a tree, a simple road map, etc., provide proof as to how many bytes had been used to encode the image and provide photographic evidence concerning the rendering of the image – i.e. what it looked like after encoding compared with the source.

The process of selection was very complex as it required a unanimous decision on a compromise between the number of bytes used and what was considered to be an acceptable rendering which in some cases was subjective. (See Figure 6.1).

It was agreed that the Bijitec solution provided the best compromise between coding efficiency and rendering and together with the inclusion of some of the aspects from the competing organisations the best overall compromise was achieved for the final solution.

In March 2000, EMS had been included in TS 23.040. Users were then able to combine conventional SMS text with pictures, animations and sounds in a singe short message. For example, an EMS message might be a birthday cake with the tune 'happy birthday' playing and the candles flickering with a text message 'To Sian love from Joe', all within a single short message.

Any of these attributes can be predefined in the receiving mobile phone without the need for the sending mobile phone to actually send, for example, the tune 'happy birthday to you' over the air interface.

In June 2001, the images were extended to include vCard (electronic business cards) and vCalendar (electronic calendar appointments).

In March 2003, geometrical shapes such as polygons were added, known as Wireless Vector Graphics.

There were numerous other enhancements to EMS during the early 2000s, adding a compelling dimension to SMS.

Sadly, as is the case with many enhancements for SMS that have been incorporated into the specification, EMS is optional and, unless the receiving mobile supports the option, then there is little point in the sender sending an EMS. There is no mechanism in SMS to assess any optional capabilities of a receiving mobile phone. The weight of market and consumer pressures usually dictates what options are deployed in the majority of mobile phones.

The term 'enhanced messaging service' can give rise to a misconception, and in hindsight it would have been better to have been called 'enhanced messaging', as it is just a way of encoding the data field of a short message and is not a service as such.

bee and flower

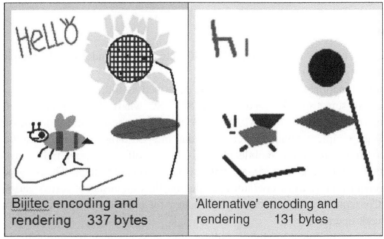

Figure 6.1 EMS object encoding example. Source: Document number T2-020210. Meeting TSG T2#16, Sophia Antipolis, 11–15 February 2002. (Extract from 3GPP TSG T2 Document T2-020210 Vector graphics comparison, page 9)

6.4 Voicemail Management[4]

The popularity and need for voicemail has shown persistent growth since the beginning of mobile telephony.

The earliest use of SMS to provide notification of 'voice messages waiting to be read' goes back to one of the first applications of SMS in 1992.

---

[4] Introduced with a Change Request to TS 23.040 in 3GPP document TP-040096.

In 2004, Research in Motion had developed a more versatile and informative Enhanced Voicemail Information system using SMS.

Enhanced Voicemail Information allows a voicemail system to convey comprehensive information to a mobile subscriber regarding individual voicemail messages and mailbox status. The information aspects are in two parts:

- Enhanced Voicemail Notification conveys information to the mobile phone regarding newly deposited voicemail messages and Voice Mailbox status such as how many messages, when they were deposited, who they are from (Calling Line Identity), etc.
- Enhanced Voicemail Delete Confirmation allows a mobile phone to maintain Voice Mailbox status information synchronisation between the mobile phone and the Voice Mailbox in the event of voicemail message deletion.

Although this feature is an option, it has no impact on interworking between one mobile phone and another, and so, if a voicemail system offers the Enhanced Voicemail Information facility, then the only requirement is for the subscriber to obtain a mobile phone that supports it.

6.5   Routers[5]

The architecture of SMS interposes an SMS-SC between the sender and the recipient which is essentially a 'store-and-forward' function.

The general functionality of an SMS-SC is described as follows.

Messages are sent from the sender to the SMS-SC where they are stored and the link with the sender terminated. The SMS-SC then attempts to deliver the short message to the recipient. If the recipient's phone is turned on and in good radio coverage, then the message is delivered. If the receiving mobile phone is turned off or in poor radio coverage, the delivery attempt fails and the SMS-SC will try again sometime later. The retry periods are SMS-SC vendor and network operator specific and thus inconsistent across various mobile networks. Consequently, the perceived performance by a user may be seen to vary according to the network operator.

In the event of prolonged delivery failures, the message may be automatically deleted by the SMS-SC after the retention time set by the sender has expired or the reason for non-delivery causes the SMS-SC to delete it. SMS allows the sender to receive delivery confirmation if requested at the time of sending the message. The absence of a delivery confirmation (if requested) means that the message has not reached its final destination, and so the sender may choose after a period of time to delete the message from the SMS-SC if such a feature is offered. Usually, for fixed network protocols such a facility is provided, but for mobile phones it is an option.

The growth of SMS applications identified that, for some applications, a message whose delivery was delayed was no longer required. If it were delivered, it could cause confusion if it were no longer relevant.

---

[5] Introduced with a Change Request to TS 23.040 in 3GPP document CP-070145.

Take, for example, a parcel pick-up service. A vehicle receiving a delayed short message may well have driven well away from a pick-up point whose location was conveyed in the delayed short message.

It became clear that SMS needed to provide a service that exhibited characteristics of 'forward' and 'not store'.

Each message in the SMS-SC has assigned to it, by the sender or by default, a validity period that defines the time a message will be kept in SMS-SC pending delivery. For some applications, the minimum validity period value was considered too long, and so messages could be held in the SMS-SC longer than the application required them to be. Conventional SMS-SCs were unable to satisfy the 'deliver and, if not, delete' requirement.

Other commercial requirements had emerged relating to behaviour in the event of non-delivery of a short message. One of these was to forward a short message to an alternative destination that *was* available. For such a requirement it should be remembered that short messages are stored in the senders' SMS-SC and that there are inevitably various SMS-SCs belonging to various mobile operators in various countries. A forwarding function would require every SMS-SC to be aware of the new chosen destination for each short message, which is impractical.

A solution was therefore needed to intercept the routing information for a short message in the mobile-terminated leg. That is fundamentally the function of the router.

The router intercepts the routing information, which an SMS-SC needs to deliver a short message, and can change that routing.

The router was adopted into 3GPP TS 23.040 in 2006 and now, along with the Virtual Mobile Equipment (VME) described in Chapter 7 section 7.1 provides considerable versatility for SMS applications. A UK company called Telsis has been particularly active in the development of routers and the VME.

6.6 Language Tables[6]

At the outset of SMS, the 7-bit default alphabet table, which allows 160 characters to be sent within the 140 octets of a single short message, was devised to accommodate some of the national variants of the countries represented in GSM at that time.

The popularity and growth of SMS penetrated countries whose characters were not wholly or even partially included in the 7-bit default alphabet. The 7-bit default alphabet table in itself had no capacity to include additional characters, and so those countries were advised to use UCS2 (Unicode), which was a 16-bit character encoding specification that catered for nearly all worldwide language characters.

Countries that were forced to use UCS2 felt disadvantaged because, instead of being able to send 160 characters in a single text message, they would be limited to 70 characters (half of 140 octets owing to the 16-bit encoding). Between 1999 and 2008 there were approximately six other countries who felt disadvantaged and raised their concerns.

---

[6] Introduced with a Change Request to TS 23.040 in 3GPP document CP-080137, and a Change Request to TS 23.038 in 3GPP document CP-080223.

In every case, 3GPP discussed their plight and advised them to use UCS2 and to adjust their SMS tariff to reflect their reduced-capacity short messages compared to those countries that could use the 7-bit default alphabet.

The matter came to a head in 2008 when the Turkish regulatory administration expressed concern about being disadvantaged and took a hard line concerning their dissatisfaction about the implications of using UCS2.

3GPP once again debated the issue at great length and realised that the ongoing increased penetration of SMS into countries that would feel similarly disadvantaged to Turkey would continue. A previous constraint of the cost of memory in the mobile phone was no longer the significant problem it used to be in the 1990s.

3GPP took the decision to provide a solution for additional country-specific tables.

The concept of escaping to an additional table for characters of international significance such as the email 'backslash' (which was not in the 7-bit default table) had long existed. The Euro symbol was added to that same extension table in 1998. However, that table had been reserved for characters of international significance and had inadequate capacity to include the various and numerous additional characters from other countries.

The solution to the problem was to extend the concept of the international character extension table to language-specific tables.

For some countries only a few characters were needed, as most of them were already contained in the 7-bit default alphabet table and so a mechanism to escape from the 7-bit default alphabet table to a specific language table for a one-shot specific character was defined, based on the escape mechanism in the international character table. It is known as the 'single shift' mechanism – analogous to the temporary shift key on a keyboard. The number of occurrences of any specific character in that language carries an overhead of an additional character, and so, while 160 characters would never be achieved, a number approaching 160 was possible.

However, the language tables also had to cater for those languages whose characters were a rare occurrence in the 7-bit default alphabet table. A 'locking shift' mechanism was defined – analogous to the locking shift key on a keyboard – to escape virtually permanently to a table containing the majority of characters used in that particular language.

As is the case for much of 3GPP standards work, consideration has to be given to backwards compatibility and the behaviour of legacy mobile phones.

Firstly, it was mandated that a legacy mobile phone receiving a short message encoded using the new language tables should not malfunction. The inability to display a message in an intelligible readable form is not considered to be a malfunction. Secondly, it was mandated that, where the single shift mechanism was used to encode a certain character, the legacy mobile should display the nearest displayable readable equivalent character or symbol from the default alphabet. So, for example, the Turkish g should be displayed as g.

However, as recently as 2009, Indian language character tables (using both the single shift and locking shift mechanisms) were added, and for many of those characters encoded using single shift there were no 'near equivalent characters' in the default alphabet table.

3GPP realised that, for non-Latin-based characters, the requirement to display 'near equivalents' was impractical, and that, faced with the probability of requirements to include language tables from other non-Latin-based countries, an exception had to be made, but wherever possible the initial rule to display near equivalent characters or symbols should still apply.

As with other SMS features, language tables are optional, and, for them to be of true international value, vendors are faced with having to implement all language tables in every phone. However, it is possible that phones for a specific country will emerge as a cost-effective compromise.

6.7 Other Important Standards Work for SMS

In addition to some of the key enhancements described above, there have been a number of important corrections without which there may have been serious issues arising concerning interworking between mobile phones of different manufacture and established SMS services.

6.7.1 Reserved Code Points

There are, in the 3GPP standards, a number of code points reserved for future enhancements and assignment by 3GPP. Initially, the term 'reserved' was taken by some organisations to mean that they were not used, and so a number of them were used for proprietary means without consultation with 3GPP.

Left unchecked, there would have been serious implications for future assignment by ETSI or 3GPP.

Additionally, it was identified that the behaviour of a mobile receiving a code point that was reserved or unsupported required clarification.

In the early 2000s, 3GPP T2 carried out a lengthy exercise examining all the unused and reserved code points in TS 23.040 and TS 23.038. The behaviour of mobile phones concerning the receipt of an unused or reserved code point was clarified, and the use of 'reserved' and 'unused' code points was unambiguously stated to be for assignment by 3GPP.

6.7.2 Port Numbers

Around September 2001, the WAP forum requested T2 to assign SMS application port numbers for WAP SMS applications. T2 subsequently assigned a number of 8- and 16-bit port addresses to the WAP Forum.

6.7.3 Support of SMS in GPRS and UMTS

During the late 1990s, the need for higher-efficiency radio bearers and higher data rates resulted in the evolution towards packet-switched networks (the General Packet Radio

Service) and consequent progress towards convergence with the Internet through UMTS (Universal Mobile Telecommunication System).

It is way beyond the scope of this publication to describe the detail of either GPRS or UMTS.

The implication for SMS was that SMS was required to be supported over a GPRS bearer without affecting the functionality of the protocol layers of SMS as defined in 3GPP TS 23.040 from the lowest layer to the highest.

The changes to 3GPP TS 23.040 have been primarily concerned with identifying new node names such as the GGSN (Gateway GPRS Support Node) and the SGSN (Serving GPRS Support Node) in the text and diagrams.

Over recent years there has been much work done on progressing the support of SMS in an IP environment and interworking with circuit-switched environments.

The underlying requirement is that SMS functionality as defined in 3GPP TS 23.040 should continue to operate unaffected.

6.8   The End of an Era

By 2004, T2's work on SMS had become primarily a maintenance task. Very few new features were being proposed – possibly owing to the fact that there was a belief by some major companies that the new media-rich multimedia messaging service (MMS) was set to replace SMS. How wrong they were!

As the pressure grew on companies to reduce travel expenditure, it was decided to move the work of T1, T2 and T3 to other groups. In February 2005, T2 held its final meeting in Sophia Antipolis. The T2 officials at this final meeting are pictured in Figure 6.2.

The documentation of all meetings in the period from 1997 to 2009 is included in the ETSI and 3GPP documentation on the respective servers (www.etsi.org or www.3gpp.org). There is also a list of all participants and lists of all documents.

TSG T, to which working groups T1, T2 and T3 reported, was dissolved. The responsibility for 'terminal-specific work' was to be merged with the CN (Core Networks) group, and a combined T and CN group was formed, known as CT. The initial plan was to create a working group (responsible for T2's work) called CT2, reporting to CT along with other CT working groups. However, in view of the fact that much of T2's work had become primarily a maintenance role, it was decided to merge T2's work into CT1 (previously CN1). The work of CT1 was dominated by the work of the previous CN1 group (which had always been responsible for core network signalling), and only two or three ex-T2 delegates attended CT1. In order to ensure that the expertise of T2 was not lost, a group of specialists from the old T2 group was established, which to this day has maintained a core of expertise dating back to the origins of SMS in the late 1980s. It is somewhat ironic that a service such as SMS has so few surviving experts from the pioneering days. The author of this chapter and the author of Chapter 5 still regularly communicate whenever issues on SMS arise. Both are considered to be worldwide experts on SMS and are regularly consulted, and their involvement as pioneers of SMS go back to its origins in standards and the industry.

Figure 6.2 The last 3GPP TSG T2 officials. From left to right: Friedhelm Rodermund, (T2 secretary); Paul Voskar, (vice chair) (Nokia); Ian Harris, (chairman) (Research in Motion); Nicola Vote, (vice chair) (NTT DoCoMo); Seung Don Han, (temporary secretary)

6.9   Further Reading

The 3GPP specifications contain technically detailed information on all of the above features. However, for those preferring a readable understanding of SMS without being overwhelmed by 'standards jargon', an excellent publication is recommended:

Gwenael Le Bodic (2005), *Mobile Messaging Technologies and Services: SMS, EMS and MMS*, 2nd edition, John Wiley & Sons, Ltd, Chichester, UK.[7]

---

[7] http://eu.wiley.com/WileyCDA/WileyTitle/productCd-0470011432.html

7

Early Commercial Applications and Operational Aspects

I. Harris
Research In Motion

One of the earliest commercial SMS applications, developed by the UK operator Vodafone in 1992, was to inform a mobile subscriber by an SMS message that there were voice messages waiting to be read on the subscriber's voice message mailbox. The voice messaging platform had a direct automated connection to the SMS-SC and a subscriber could elect by prior subscription agreement with Vodafone to be alerted of voicemail messages by either a mobile-terminated voice call or by a mobile-terminated short message. Whilst, at the time, voicemail notification by SMS seemed a useful and innovative invention, delayed short messages could result in the mobile subscriber being told that messages were waiting when in fact he had already read them.

The preference for many subscribers fell back to being notified of their voice mailbox messages and voice mailbox status by voice.

Another useful early application developed by Kevin Holley while working for the UK operator Cellnet was to extract news items from the Internet and convert the content to simple text short messages.

For many years, the use of SMS for commercial applications was constrained by the impracticality of keying in messages by hand on a mobile phone.

In the mid-1990s, two important developments took place that would stimulate the growth of commercial applications:

- provision of fixed-network connections to SMS-SCs by third parties;
- provision for the connection of intelligent terminals to the mobile phone.

Short Message Service (SMS) Edited by Friedhelm Hillebrand
© 2010 John Wiley & Sons, Ltd

7.1 Fixed-network Connection to the SMS-SC

The GSM SMS specification TS 03.40 had, from the beginning, an architectural boundary that specified that any connection to the SMS-SC to any entity other than a mobile telephone network was outside the scope of the standard. That situation remains true today.

Many business applications require fixed-network connectivity to an SMS-SC because of the impracticality of using a mobile phone. There was and still is no common agreement between network operators on the type of fixed network, the physical connectivity or the protocols that should be used to connect SMS-SCs to a fixed network. Because there were a number of vendors for SMS-SCs, each vendor developed its own proprietary interface to fixed networks. and mobile network operators had their own preference for one particular SMS-SC vendor.

Third parties wishing to offer commercial services across all mobile networks were faced with having to provide connectivity to every network operator's SMS-SC, which entailed different physical connections and different protocols.

The problem was brought to the attention of ETSI in 1995, and much debate followed as to whether the physical connection to the fixed network and its protocol should be standardised by either selecting one of the proprietary protocols or by developing a new one containing all the features of the proprietary protocols.

Common agreement was never reached by mobile network operators and their SMS-SC vendor to adopt a protocol of one of their competitors, for reasons of loss of commercial status but also because the level of sophistication of the various protocols varied and some were unsuitable for some applications. Considerable costs for re-engineering hardware and software changes would also have been required to adopt a different protocol.

SMG4 concluded that, as there were already various commercial applications operational, a new ETSI protocol standard would be a waste of time and effort.

However, in an attempt to inject some stability into fixed-network connectivity and restrict the proliferation of protocols, ETSI published a technical report TR 03.39 that contained the protocol details from various SMS-SC vendors and advice to future SMS-SC developers to adopt one of the existing protocols rather than invent a new one. The proprietary SMS-SC to SME interface protocols were contained in the annexes of TR 03.39 as follows:

- Annex A: Short Message Peer-to-Peer (SMPP) Interface Specification (Aldiscon Information Systems).
- Annex B: Short Message Service Centre External Machine Interface (Computer Management Group).
- Annex C: SMSC-to-SME Interface Specification (Nokia Cellular Systems).
- Annex D: SMSC Open Interface Specification (SEMA Group).
- Annex E: SMSC Computer Access Service and Protocol Manual (Ericsson).

In the transition from ETSI GSM to 3GPP, TR 03.39 was renumbered TR 23.039. Maintaining and updating the detail of each protocol in TR 23.039 as SMS-SC vendors improved their products became an onerous task, and TR 23.039 was modified to remove the protocol detail and provide a reference to each of the SMS-SC vendors only. Even that proved difficult to maintain as company's merged, reorganised or were dissolved, and so, in 2002, after Release 5, TR 23.039 was withdrawn.

The numerous proprietary fixed-network protocols were to have a significant negative impact on rolling out commercial applications across various mobile networks, particularly for those applications that required two-way messaging and for those that required one-way mobile-originated messaging.

In order to understand the technical and commercial issues surrounding this, it is first important to understand some fundamental aspects of the function of the short message service with regard to addressing and routing short messages.

A message originating from a mobile phone normally goes to the SMS-SC that belongs to the originator's network operator. This is primarily because a network operator has no means of billing a subscriber that does not belong to its network. The message is held in that SMS-SC, from where delivery attempts to the recipient are made irrespective of the network operator to which the recipient belongs. Messages are never sent from one SMS-SC to another SMS-SC. They are stored in the SMS-SC of the sender's mobile operator's domain until they are either delivered or they expire (or are deleted).

If the recipient of a message is in a different network operator's domain to the sender of the message, then any reply to that message will normally be sent via the replying subscriber's SMS-SC and not via the SMS-SC that sent the original message. There is, however, provision to route a reply message back via the original sender's SMS-SC. That is called Reply Path. However, not all SMS-SCs support Reply Path, and so the operation is not guaranteed. Reply Path allows originators to indicate to the SMS-SC that they are prepared to pay for the reply message. Reply Path is described in Section 7.5.

A commercial application that required just mobile-terminated messages was able to operate on all mobile networks because it was only necessary for the commercial application to connect to one network operator's SMS-SC. That SMS-SC would handle the mobile-terminated message in the same way as any other message it held destined for a mobile on any network.

However, where there was the need in the commercial application for a mobile-originated message to be received from any network operator's subscriber, then connectivity of the commercial application to every network operator's SMS-SC was necessary, as explained above.

Take, for example, a bank wishing to offer a banking service to its customers that requires two-way messaging with a mobile phone that could be on any mobile network.

For fixed-network-originated messages, the bank's application server would be able to send messages to all its subscribers no matter which network operator they belonged to by just connecting the host server to one network operator's SMS-SC. However, messages sent from a mobile either as an originated message or as a reply message could only reach the bank's application server if the latter was connected to *every* SMSC in *every* network operator's domain.

The problem remained unresolved for many years until, in the year 2000, the author of this chapter, whilst working for Vodafone, identified a solution that was amazingly simple.

The solution identified the need for an intelligent platform in the network that would behave like a number of mobile phones. Vodafone called the platform the VME (Virtual Mobile Equipment).

A mobile subscriber wishing to access the commercial application would merely send a short message to what seemed to be another mobile phone number. Alternatively, the mobile subscriber could use a published 'short code' (e.g. 1234) to send the short message, which was then translated to a mobile telephone number by the sender's SMS-SC. Consequently, no matter what mobile network the subscriber was on, as with a normal short message being sent to a real mobile in another network, the message would find its way into the mobile network to which the number belonged. The network with the VME was where the message would be routed to. The VME then translated the destination address into an address that the SMS-SC would recognise as being a fixed-network address and not a mobile phone. The message could therefore be routed to the application server irrespective of its origin.

The benefit was that the commercial application server only needed to be connected to one network operator's SMS-SC that contained the VME.

This provided a solution to the growing interest for using SMS to vote interactively with TV programmes.

The use of the VME concept is in wide use today and is covered under the subject of Routers in TS 23.040.

There were also a number of simple SMS-SC access mechanisms whereby simple data terminal type devices could construct short messages and send them to mobile phones via, for example, 'dial-up modems'. One such gateway was developed by Kevin Holley while working for the UK operator Cellnet.

7.2   Network Operator Interworking, Roaming and Number Portability

It is hard today to believe that up until the late 1990s it was not generally possible for a subscriber of one network to exchange short messages with a subscriber of another network (network operator interworking), although they could exchange voice messages. Additionally, although subscribers could use their mobile phone for voice on any network, they could not send short messages when roaming to another network.

The root of both of these problems was primarily commercial, but there was a technical deficiency in the way that SMS messages were handled compared to speech, which meant that no commercial solution was possible without technical changes.

Ironically, these problems only existed between network operators who were in competition with each other, and so, when roaming abroad to a 'friendly' network operator, subscribers could use SMS as though they were on their home network.

In the UK, subscribers were becoming increasingly frustrated, not understanding why voice was not a problem but SMS was. OFTEL (UK Office of TELecommunications Regulator) insisted that the problem was resolved.

To compound the issue, there was an increasing demand from mobile subscribers to transfer their subscription to a different mobile operator but to be able to retain or 'port' their mobile number to the new network, including the NNG (the National Group Number part of a mobile number, traditionally identifying the network to which that mobile belongs). Again in the UK, OFTEL insisted on a solution.

Technically, all three problems were concerned with identifying the network to which the sender of a short message belonged and taking action concerning the routing of the mobile originated short message.

There was no means by which a network operator could bill the sender of a short message who was not a subscriber to that network, and so any short message originating on that network from a subscriber who was not subscribed to that network would be rejected by that network's SMS-SC. The NNG was the means of identifying whether or not an SMS-SC would accept a mobile-originated short message. Furthermore, the network to which the subscriber had roamed had no means of sending the mobile-originated short message to the subscriber's home network SMS-SC.

The basic requirement for an SMS-SC to accept SMS messages only from senders who belonged to that network (or a friendly network) was a principle fundamental to billing and had to be retained. That principle still exists today.

The solution to all three problems required technical changes along the following lines.

Every mobile network SMS-SC was required to carry out additional checks to allow routing to an alternative network for short messages originating on their network from subscribers of another network. The basic check of the NNG already in place to decide on acceptance or rejection of a short message was elaborated. If the NNG belonged to that same network, then the short message was allowed into the SMS-SC. If not, then the NNG would be used to route the short message to the SMS-SC in the network to which the sender belonged. However, before deciding to route the short message to another network, an additional check was made to see if the sender had ported to the network on which the short message had originated even though the NNG identified the sender as potentially being a subscriber to another network. If the sender was ported, then the message was allowed into the SMS-SC. If the sender was not ported to that network, then the short message would be routed to the SMS-SC to which the NNG belonged, where the check for a ported number would again take place.

7.3   Third-party SMS-SCs

From the outset it was assumed that the SMS-SC would be under the strict control of the mobile network operator. This was primarily because the SMS-SC was required to connect into the mobile network signalling system (CCITT No. 7), which also gave access to sensitive areas within the operator's domain such as the HLR (Home Location Register) which holds subscription details of its subscribers.

In the mid-1990s there was a general growth of liberalisation for mobile services, but, for network security reasons, the vast majority of mobile network operators saw the need to retain control of the SMS-SC, their view being supported by the regulators.

However, in the interest of fair competition, some operators such as Libertel in Holland decided to allow third-party SMS-SCs to be developed and connected to their network which were not under the direct control of the network operator.

In a few cases, in some countries where control over the connection and supervision of third-party SMS-SCs was not as tight as it should have been, it allowed SMS spam to develop. It only requires one such SMS-SC to exist to allow SMS spam to be sent to every mobile subscriber in every network.

Where that has occurred, mobile network operators have had to block all SMS traffic from the network containing the rogue SMS-SC, hence curtailing legitimate use as a consequence.

Clearly, spam can be sent from any device capable of sending a short message, such as a mobile phone, but it is more difficult to target large numbers of recipients from a mobile phone and billing usually controls such traffic.

Most mobile network operators procure their SMS-SC from one supplier, as there are many design and performance aspects of an SMS-SC that are outside the scope of 3GPP standards but specific to the business offering from a particular mobile network operator. There may also be a direct bearing on the quality of service offered by a network operator where there are multiple vendors, and it is essential that a network operator is able to provide a comparable quality and level of service.

A third-party SMS-SC is somewhat outside the network operator's control, and so users may experience inconsistent levels of service quality for which they often see the network operator as being responsible.

Any SMS-SC enhancements that affect the network operation of SMS have to be incorporated into every SMS-SC. An example of this is the solution to the network operator interworking, roaming and number portability problem described earlier. Where there are third-party SMS-SCs, control over the changes to their product can be difficult, and hence there is a risk of jeopardising SMS.

Rather than allow third-party SMS-SCs, it is far better to encourage the development of applications by third parties and connect them to the SMS-SC that is under the direct control of the network operator. Such is the situation in the UK and most other countries.

7.4 Intelligent Terminal Connections to Mobile Phones

Some early SMS commercial applications required client software intelligence at the mobile phone. Clearly, it was impractical to embed such software for the numerous applications in the mobile phone, as every phone would have to be specific for an application or engineered for a choice of applications.

One of the earliest of these applications was for vehicle tracking (e.g. parcel delivery/collection). Vehicles fitted with GPS were required to send their position periodically via SMS, which necessitated an interface to be developed to a mobile phone to allow the attachment of an external intelligent programmable device such as the Psion organiser or the Palm. For some applications, bespoke hardware was developed.

Because there was no standard interface or protocol, the matter was brought to the attention of ETSI SMG4, who felt that an external interface to a mobile phone should be standardised. ETSI SMG4 created TS 07.05 (see Chapter 5), which defined three alternative protocols operating over an asynchronous V24 interface:

- Block mode;

- Text mode;
- PDU mode.

The Block mode is a binary protocol and is suitable for software-driven applications. It includes error protection and is suitable where there is a risk of interference between the application terminal and the mobile phone.

The Text mode is an AT-command-based protocol and is suitable for a human interface and has no error protection.

The PDU (Protocol Data Unit) mode is a character-based but hex-encoded protocol suitable for software-driven applications and uses AT commands.

The combination of an intelligent terminal connected to the mobile phone and fixed-network connectivity opened up the opportunity for many commercial opportunities. One such application is vehicle tracking for parcel collection and delivery.

7.5   SMS Keyboard Text Entry

The drive for smaller mobile phones into the late 1990s and hence smaller keypads made SMS text entry increasingly difficult. Given the dramatic increase in the use of SMS, attention turned to easing the task of entering short message text.

One particular company called Tegic, developed a text predictive algorithm whereby software would predict from the start of a word string what the word was likely to be and thus save the effort of entering characters. Tegic brought their idea to ETSI SMG4, but SMG4 felt that this was beyond the scope of standards as it was an implementation matter. Tegic embarked on a campaign to lobby mobile vendors and eventually succeeded in making a significant market penetration. Today there are several predictive algorithm implementations – some mobile vendor specific.

By the late 1990s, SMS was becoming increasingly popular among the younger population. Many youngsters were already familiar with Internet chat rooms where the use of 'shorthand text' such as CUL8TR ('see you later') was commonplace. The 'shorthand text' found its way into SMS messaging. Probably unknowingly, they had also stumbled on an incidental means of keeping the number of characters in a short message to a minimum and in so doing avoided higher charges than they may have incurred for concatenated messages.

The 'fun' attraction of SMS using 'shorthand text' was a major factor in the explosion of SMS text messaging towards the end of the 1990s.

7.6   SMS to Fax and SMS to Email

The most significant growth of SMS in the mid- to late 1990s was conventional subscriber mobile-to-mobile messaging. However, it was realised that there were other established technologies such as fax and email that were capable of terminating the data content of a short message, in that both technologies were capable of displaying text.

Vodafone developed an SMS-to-fax service that provided one-way messaging from a mobile phone to a fax machine. Sending from a fax machine to SMS was never pursued, as it required a complex scanning process to convert facsimile to text. Commercially, the success of the SMS-to-fax service was somewhat limited, as the popularity of fax began to give way to the use of email in both mobile and private networks for versatile two-way messaging.

Email seems ideally suited to interworking with SMS. However, there are some fundamental differences, as will be explained.

Short messages predominantly use destination addresses that are numeric, and, although there is provision in SMS addressing to use alphanumeric addresses, the length of such an address is limited and cannot contain non-alphanumeric characters. Internet addresses are of the form lucy@wondernet.com, and SMS cannot handle such an address in its destination address field.

Much work was done in TS 03.40 in the late 1990s to make provision in the mobile phone to indicate to the SMS-SC that a message was to be sent to an email address. This is described more fully in Chapter 5.

The intended destination email address had to be embedded into the 160 character user data field of the short message. The format of the email address and its position in the body of the short message had to be explicit, otherwise the SMS-SC would not be able to discriminate between normal text and an Internet address. The email address also consumed a portion of the available space for short message text, and so less than 160 characters were achievable per single short message.

For concatenated short messages, the email address had to be present in every short message segment because it is possible for segments to go missing.

For mobile-terminated messages sent from an email source, it is necessary for some intelligence attached to the Internet to realise that the destination address is actually intended for a mobile phone, and so translation of the Internet address to a mobile telephone number is necessary.

Given the above complications of email interworking, it is not surprising that the service has had a limited success.

Around 1995, Vodafone commenced a project for two-way interworking between SMS and email. Other companies such as SendIt, Psion and Purple Software had also been involved in similar projects, and so Vodafone worked jointly with these other companies to develop an SMS-to-email service known as Vodafone:m@il. As part of that project, the need for an additional specific code point in the TS 03.40 Protocol Identifier value arose and was adopted into TS 03.40 in May 1995.

The Vodafone:m@il project brought a commercial reality to the fundamental differences between SMS and email:

- *Incompatible addressing.* SMS is a numeric-based addressing system, whereas email is a name-based system.
- *Content discrepancies.* SMS is basically limited to 160 character messages, whereas email has almost no bounds on message size. SMS, through EMS, is capable of handling simple graphics animations and sounds, whereas email can handle very complex images and sounds.

Figure 7.1 Souvenirs of Vodafone:m@il

- *Tariffing.* SMS messages are chargeable, whereas email messages are ostensibly free. SMS messages are normally charged to the sending party. It is difficult for the mobile-terminated leg of a short message to be charged to the sender because there is no contractual link or agreement with the sender, although interoperator accounting can balance message handling costs. Free email-originated messages would cause an explosion in mobile-terminated message traffic, for which there is no easy revenue path.

Vodafone:m@il was finally abandoned a few years later owing to the impracticalities described above. Some souvenirs of the project are shown in Figure 7.1.

7.7 Two-way Real-time Messaging Applications

There have been numerous applications developed that have attempted to use SMS for two-way real-time messaging.

SMS is not particularly suited to such applications because of the risk of delays usually brought about by the receiving mobile being in less than perfect radio coverage.

Messages sent to a fixed-network termination have a very high probability of delivery on the first attempt, i.e. within a few seconds of it having been sent from a mobile. Messages sent to a mobile phone, however, may experience delays of many tens of seconds or even minutes and, in extreme cases, hours. The SMS-SC retry schedules in the event of a delivery failure are outside the scope of ETSI/3GPP specifications and so will vary between different networks, making applications vary in timeliness of delivery across different networks.

Despite the 'store-and-forward' characteristic of SMS, a mobile phone in good radio coverage can give the appearance of a real-time two-way messaging service. However, mobile-terminated short messages are, in general, subject to a higher delivery delay than mobile-originated short messages terminating at a fixed-network entity. This is primarily due to the receiving mobile being in less than ideal radio coverage, which can result in non-delivery on the first attempt and sometimes non-delivery on subsequent re-attempts. In certain cases, mobile-terminated short messages may never be delivered. Hence, for services relying on two-way real-time SMS messaging, the performance cannot be guaranteed.

One two-way service that has recently gone into commercial service in 2009 in the UK is known as 'eSMS'. This is an emergency SMS service for the hearing impaired. However, it relies on human intervention at a central point to convert the mobile originated short message text into voice for onward routing to the emergency response service and vice versa for reply messages. Despite concerns over delays, this is a pragmatic case of something is better than nothing and the expectations of the service are made clear. The service is intended for those hearing impaired persons choosing to register for eSMS and not for use by the general public.

It has been suggested that SMS could be used for the broadcasting of public warnings such as earthquake, flood, terrorist attack, etc. Such warning messages are required to be sent to specific mobile phones in a particular geographic area.

SMS has two significant problems for such applications:

- Firstly, SMS has no inherent location information. The location of a particular mobile phone can be obtained by independent means, but the process is cumbersome and time consuming, particularly when trying to identify all mobile phones in a particular area that are on different mobile networks.
- Secondly, SMS is a 'point-to-point' service, and consequently every short message to every subscriber from each network operator would be sent sequentially, working down the list of mobile recipients, as opposed to concurrent delivery. As a consequence, there may be a significant time difference between the first mobile subscriber receiving a message and the last subscriber receiving the same message.

Far better technologies exist, such as the Cell Broadcast service (historically and originally called Short Message Service Point-to-Multipoint).

7.8 Performance

Network operators have a vested interest in monitoring the quality of service for all of their services in order to identify potential problems and to investigate problems that affect their quality of service.

Most SMS-SCs have inbuilt SMS records that show when a particular message arrived in the SMS-SC, when it was delivered, the mobile numbers of the sender and the recipient, etc.

Analysis by Vodafone of their short message service in the late 1990s gave the following results, which were, at the time, consistent with the results of other UK network operators:

- The time between a message being sent from a mobile phone to the time that message was received at the recipient's mobile was typically 6–8 seconds. Approximately 1–2 seconds of this was attributed to message storage in the SMS-SC.
- The time between a message being sent from a mobile to that same mobile receiving a delivery confirmation was typically 10–12 seconds. (Note that the indication of 'message sent' on most phones is only confirmation that the message has reached the SMS-SC and NOT that it has reached the recipient's mobile).
- Typically, 38% of messages were not delivered on the first attempt to mobile phones, primarily due to the receiving mobile being out of coverage, in poor coverage or turned off. However, 98% of all messages were delivered on the first attempt where the destination was a fixed-network termination.
- Typically, 98% of messages were eventually delivered, provided that the retention time for the short message was set to 3 days.

Although it has not been possible to obtain the latest statistics, it is unlikely that the above performance will have changed significantly although there may well have been an improvement in the 38% figure for the first delivery attempt of mobile-terminated messages owing to improved mobile radio coverage.

The SMS-SC records provided an insight into a user's experience of the short message service but could not provide quality-of-service measurements, as the source and destination devices were uncontrolled.

To address this problem, Vodafone developed an automated performance measuring system that deployed mobiles around the UK that were connected to intelligent terminals which allowed those mobile phones to report back their perception of what had happened and allowed correlation with the records of messages at the sending end.

Each day, thousands of records were analysed, giving a real-time performance indication of the behaviour of Vodafone's entire short message service. Hence, when problems were reported from subscribers, Vodafone was able to check the performance of the SMS network at that time and, if necessary, isolate the problem as being specific to the subscriber or the application.

7.9 SMS Traffic Growth

There was no precedence for specifying the message handling capacity for the first SMS-SCs that were developed around 1992. SMS was a new service whose use was unquantified.

Vodafone's SMS-SC was developed by SEMA – a company in Reading, UK, close to Vodafone's headquarters in Newbury. Because of some architectural deviations from the TS 03.40 specification, Vodafone called their SMS-SC a Text Messaging Centre (TMC). It was not foreseen that the term 'text messaging' would become a ubiquitous term for SMS.

The capacity for message handling by the TMC was based on the experience of voicemail, which, given the relatively small number of mobile users at that time, was 10 messages per second across the whole Vodafone network.

As the number of mobile phone users increased, so did the use of SMS, and during the early to mid-1990s Vodafone's TMC message handling capability increased from 10 messages per second to several 100 messages per second. By then there were about 45 billion short messages being sent per month – mainly in Europe.

In the late 1990s, SMS traffic growth was virtually exponential, serious redimensioning SMS-SC message handling capability became necessary and network resilience and redundancy were needed. SMS-SCs were designed with faster processing capable of handling many thousands of messages per second, but the limitation was not in the processor but in the connectivity to the mobile network. This necessitated SMS network topography re-thinking to allow deployment of large numbers of SMS-SCs to cope with traffic growth.

There are many facts and figures available within a network operator's domain that give SMS traffic statistics, but these are not normally publicly available. Suffice it to say that, in the current-day figures, the number of SMS messages being sent worldwide per year is measured in hundreds of billions. This is discussed more fully in Chapter 8.

7.10 Billing

The realisation in the late 1980s that SMS was destined to become an operational service resulted in a discussion among network operators on how much should be charged to subscribers and how this should be done.

There was no model of any other service on which to set a tariff.

A memorandum of understanding was set up among mobile network operators to address the issue of 'billing'. It was known as MoU BARG (Memorandum of Understanding – Billing Accounting Regulatory Group).

This group formulated agreements among mobile network operators on how SMS was to be billed and tariffed.

The initial concept was to treat SMS like any other switched-circuit voice call, i.e. the originator would be charged. The SMS-SC was in the sender's domain, and so it was a simple matter to extract information from the originator's SMS-SC concerning the identity of the subscriber who sent the short message. Only subscribers to a particular network would be allowed to send short messages to that network's SMS-SC. Other subscribers' short messages sent to that SMS-SC would be rejected.

The delivery of a short message from the SMS-SC to the recipient meant that, where the recipient was on a different network to the sender, the recipient's network was required to handle the short message. However, the recipient's network was not designed to monitor the detail of short messages other than the mobile network from where the short message had been sent, and in any case there was no commercial billing agreement between the recipient's network operator and the sender of a short message if the sender was on another network. To put it simply, senders could not be billed by anybody other than their own network operator.

The concept of inter-network accounting was developed: the number of short messages that a network received for delivery from another network operator was tallied, and an agreement was reached between them to balance the cost of handling the delivery of short messages.

Apart from the basic concept of billing the sender, arrangements were made for bulk billing – mainly for commercial operations where preferential rates applied to a sender who sent high numbers of short messages.

SMS value-added services were also developed, where premium rates for certain messages applied.

The cost of a mobile-originated short message in the UK was set at 10 pence around 1990.

Today, there are numerous packages allowing flexibility for SMS users, e.g. free SMS messages rolled in with a package agreement for voice services.

Occasionally, the question is asked whether a short message tariff should be based on the number of characters sent in a short message. The reality is that the unit size of 160 characters is fundamental to the technology of SMS, and it costs just the same for a network operator to handle a message of one character as it does to handle a message of 160 characters.

7.11 The Content Provider Access (CPA) Model Deployed in Norway[1]

Mobile network operators in Norway agreed around 2000 to adopt a model for offering content providers the opportunity to connect to the operators' short message service. Of the most important characteristics of the model, the following two should be mentioned:

1. The content provider is granted an access number to be used by the end-users for accessing their content. This number is the same, irrespective of which mobile operator holds the end-user as a customer.
2. The mobile operator bills the end-user for both air time and content, preferably in an integrated way to make it more easy for the end-user to perceive the total expenditure. The operator then transfers a certain percentage of the content fee at a later stage to the content provider. The percentage is an essential part of the deal between the content provider and the mobile operator. The operators have agreed to a common price structure.

It is a common conclusion in the mobile community in Norway that the great success of the CPA model in the Norwegian market is in particular due to the two aspects mentioned above.

CPA has been extended also to other access methods, such as WAP.

7.12 SMS in 2009

In 2009, SMS remains a significant revenue earner for mobile network operators and is as popular as ever for mobile subscribers.

---

[1] Based on input from Finn Trosby.

The attention in 3GPP standards bodies has recently been turned towards the next stage of evolution for the infrastructure of mobile telephony, LTE (Long-Term Evolution).

It is beyond the scope of this publication to enter into the detail of LTE.

What is of significance is that work has long been under way in 3GPP to adapt SMS, which was originally designed for circuit-switched bearers, to packet-switched bearers and interworking with the Internet.

As recently as August 2009, discussions were under way in 3GPP concerning SMS-only mobile phones for LTE.

SMS remains a unique service, and the drive to support SMS for LTE is a testimony to the continued popularity of SMS today among its users and its commercial viability for those responsible for providing the service.

8

Global Market Development

F. Hillebrand
Hillebrand & Partners

The widest use of SMS is for the person-to-person exchange of short messages. This chapter focuses on this aspect of SMS.

8.1 The Creation of a Large Base of Mobiles and the Global SMS Infrastructure

After the completion of the standardisation, all operators were bound to the commitment from the GSM memorandum of understanding to implement essential services such as SMS. Therefore, they included the requirement to support SMS point-to-point in the procurement activities for their networks and mobiles. This created a wave of demand for SMS equipment. Thus, every mobile station supplier and every systems supplier had to develop SMS functionalities.

The first portable telephones available in 1991 and 1992 did not support SMS. However, hand portables started to support SMS point-to-point mobile-terminated in 1993. The pioneers were Ericsson, Nokia, Orbitel and Siemens. All mobiles supported the mobile-terminated service from 1994 onwards, and the mobile-originated service from 1996 onwards.

In the GSM network, a new server, the SMS-SC, was required. The rest of the functions in the network were embedded in protocols needed for telephony (MS/BS protocols and MAP). The functions in MS/BS protocols and MAP were available in the systems of all system manufacturers from the end of 1994 onwards. For the development of SMS-SCs, a number of dedicated companies appeared, in addition to system manufacturers. SMS-SCs were available from Aldiscon Information Systems, CMG Computer Management Group, Ericsson, Nokia Cellular Systems and SEMA Group. This is a remarkably broad range of

Short Message Service (SMS) Edited by Friedhelm Hillebrand
© 2010 John Wiley & Sons, Ltd

suppliers for a totally new system element in a market that was relatively small at the time. There were some early acceptance tests already in 1992, but in most networks these tests took place in 1993–1995. The procurement and test activities led to the availability of SMS in practically all GSM networks from 1995 onwards.

International roaming was available from 1995 onwards.

National interworking for message transport between competing networks in a country lasted until 1996 (in exceptional cases, until 1997) owing to protracted commercial negotiations. When SMS could be sent between networks, this removed the last barrier to widespread acceptance.

8.2   First Use of SMS by Network Operators

In 1991, the scepticism about SMS was summed up by the director of one mobile network operator in the words: 'Why would anybody want to send one of these messages when they can talk to them?'. It is fair to say that, in 1992, when the first short message was sent, nobody foresaw that SMS would become the most important revenue earner next to speech for the following 18 years and perhaps beyond. A message service that was limited to 160 characters and that had to be keyed in on a mobile phone keypad using fingers and thumbs, and whose message delivery could not be guaranteed, was hardly likely to attract much attention.

As the new Europe-wide mobile telephony was so exciting, nobody marketed SMS in a systematic way. In 1993, some operators used SMS to notify users of received voicemails in order to trigger new calls by their customers to the persons who had left the voicemail. Other operators saw no use for it. They left the functionality in the network without promoting it and without charging for it.

8.3   How SMS Was Discovered by Young People and Became a Part of the Youth Culture and Widely Accepted

Starting in the mid-1990s, youngsters in many countries could afford mobiles. They realised that SMS was a simple and cheap way to communicate with their friends. They loved this possibility, and, as it was not known to grown-ups, messages could be sent secretly; for instance, underneath the desk at school, e.g. C U 2 Nite ('see you tonight') or C U L8R ('see you later'). This SMS language shorthand was derived from abbreviations used in Internet chat rooms. It was used to economise on the use of characters. It made SMS 'fun' and created a community feeling. Young people developed an admirable skill to input SMS characters by the standard telephone keypads. Thus, neither the limitation of 160 characters nor the clumsy inputting method was disadvantageous to SMS development.

As SMS was available everywhere and in every mobile, it became part of the youth culture in all GSM countries. In many countries a substantial part of the total SMS revenue comes from young people using the service as a message and chat carrier. This started back when the user interface for SMS was quite poor. There seems to be light years of distance between the skills and interest in SMS of a youngster (12–20 years old) and those of a lady above 50. Unexpectedly large traffic volumes appeared in all networks. Those network operators who had offered SMS free of charge in the beginning had difficulties in getting their customers to accept charging.

Today, SMS point-to-point services are used very widely by people of all ages. Grandparents were persuaded by their grandchildren who had received a mobile as a gift to exchange messages with them.

The high hopes that SMS as a value-added service (VAS) would attract a lot of competing VAS operators have not become reality.

8.4 SMS Has Become the Leading Mobile Messaging Service and Will Stay in the Lead in the Foreseeable Future

8.4.1 Global SMS Traffic and Revenues in 2008

It is not possible to obtain consolidated global figures of the total message number, as not all operators report their figures. Therefore, three different approaches will be presented.

8.4.1.1 Model 1: a Simple Calculation

Table 8.1 presents figures on the average number of messages per user and month.[1]

This allows a cautious estimate of the total traffic and revenue:

- *Traffic*: 3.55 billion users × 70 messages per user and month results in 250 billion short messages per month. This results in 3 trillion short messages per year.
- *Revenue*: 3 trillion short messages per year × €0.03 per short message[2] results in a revenue of €90 billion per year.

Table 8.1 Average number of messages per subscriber

Operator in areas	Messages per subscriber and month in Q4 2008
Industrialised developed countries	40–100
Special situation in developing countries	170–950

[1] Derived from a table of Wireless Intelligence.

[2] This low figure takes into consideration bulk rates for large customers, flat rates, etc.

8.4.1.2   Model 2: Research by a Market Research Firm

Portio Research estimates $US 89 billion SMS revenue for 2008 generated by 3.5 trillion short messages in 2008. They forecast a rise to $US 100 billion in 2010 and 5 trillion short messages in 2011.[3]

8.4.1.3   Model 3: Another Crude Extrapolation

The 2008 UK SMS (source MDA) was 78.9 billion short messages for the year.[4] The Population Bureau, Washington, estimated the UK population in July 2008 to be 62 million, and the world population to be 6.7 billion.[5] Assuming 100% penetration in the UK, this would result in a potential global text market of:

$$(79 \text{ billion short messages in UK per } 62 \text{ million users in UK}) \times$$
$$3.55 \text{ billion users worldwide} = 4.5 \text{ trillion short messages worldwide}$$

8.4.1.4   Conclusion

It can be estimated that the world market saw in 2008 about 3–4 trillion short messages and a turnover of the order of $ 80–100 billion.

8.4.2   Some Considerations about the Market Development

Compared to the expectations at the time of the standardisation, there were major changes in the mobile communication market. To begin with, car telephones (which no longer exist) were in the prominent position. But customers have for years been using only personal hand-portable phones. This has strengthened the character of SMS as a personal messaging service.

The very high penetration of GSM phones in many markets and the acceptance of the GSM standard in nearly all countries of the world have made SMS, which is integrated in GSM, a ubiquitous service. These developments have increased SMS usage beyond all expectations one might have had at the time SMS was standardised.

In the meantime, new mobile messaging services have appeared in the market: MMS (Mobile Multimedia Messaging Service), Mobile Email, Mobile IM (Instant Messaging). The emergence of MMS in 2005 was predicted by some to be the demise of SMS. In reality, that has not been the case. Unlike SMS, MMS offers the capability of sending complex graphics and media-rich messages, and yet its penetration has been somewhat stunted.

---

[3] www.portioresearch.com, new study 'Mobile Messaging Futures 2009–2013'.
[4] www.text.it
[5] http://www.prb.org/pdf08/08WPDS_Eng.pdf

8.4.3   A Forecast for 2013

Portio Research have estimated the revenue in $US billion for the different mobile messaging services in 2008 and 2013 (Figure 8.1).[6]

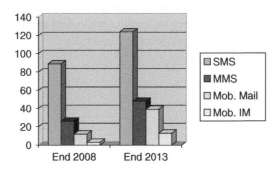

Figure 8.1   Revenue for different mobile messaging services worldwide

This diagram shows that SMS is now, and will remain for the foreseeable future, the dominant mobile messaging service.

8.4.4   Long-term Prospects for SMS

Another indicator in this direction is that the hottest and fastest-growing messaging service on the Internet is not a multimedia-rich service but Twitter, a type of broadcast SMS with just 140 characters per message.

SMS still has enormous scope for traffic growth. Whereas person-to-person SMS is widely accepted by young people, there is much room for use by older people. The fear of many, especially among mobile operators, that IM and email might cannibalise SMS is greatly exaggerated. SMS has the great advantage over IM and email to be available on all mobile phones. SMS is installed in 4 billion phones worldwide. IM and email rely on a part of the mobile phones equipped for these services and on 1 billion personal computers in service in June 2008 which can be configured for IM and email.[7]

However, as mobile email becomes more established in the mobile workplace, it is inevitable that, as the cost of mobiles that are capable of email decreases, the attraction of SMS will be reduced, and the versatility of email may well begin to have an impact on SMS. However, it remains an open question as to whether mobile email will play a larger role in private use and in the large volumes of very cheap mobiles used in developing countries or in the lower end of the market in developed countries. Only time will tell.

---

[6] Graphics designed on the basis of numbers provided by Portio Research in the leaflet for their new study mentioned in the previous footnote.

[7] Wikipedia, see 'Personal Computer'. The shipments per year were 1 billion mobiles, compared with 264 million computers in 2007.

However, consideration is now being given in standardisation and in large companies to using SMS as the transport mechanism for machine-to-machine communication (m2m). This means that every car, every machine in every factory and even electricity, water and gas consumption meters could in the future be equipped with SMS-capable mobiles.

The reality is that, in 2009, SMS is as popular as ever and shows no sign of diminishing. SMS is enjoying a long history of success analogous to the history of telex and facsimile – all services attributing their success to simplicity.

9

Conclusions

F. Hillebrand
Hillebrand & Partners

9.1 Factors that Were Critical for the Success of SMS

SMS point-to-point is popular because of its simplicity as a personal messaging service. Many customers carry their phones with them all the time and are normally able to send short messages wherever they are (at home, in a neighbouring country or far away), and whenever they want, to every GSM/UMTS phone in the world. SMS is the ideal messaging service for short notices. It normally reaches the addressees wherever they are, and it transfers the information as soon as possible. If the addressee's phone is switched off or outside radio coverage of a network, then the short message is normally received as soon as it is possible for the network to deliver it. The received messages are normally stored in the recipient's phone and can be read at a convenient time. Thus, the sender need not to be concerned about whether the recipient is reachable at the time of sending the short message. The use of SMS by both sender and recipient is far less intrusive and hence socially more acceptable than voice calls. And it outstrips any messaging service based upon the fixed networks – e.g. email – as people tend to carry their handsets all day, whereas they only sporadically sit down at their fixed-network terminal.

Customers value the price they have to pay for the service they get as fair. In the early years of the GSM commercial service, a short message was cheaper than a phone call. But today customers love the convenience of SMS, even if it is more expensive than the shortest call in the cheapest tariff.

SMS is popular among network operators, as it can be implemented with a limited additional investment and is very profitable. However, it remains to be seen what the impact of the recent very deep price regulation in Europe will be on the ability of operators to enhance capacities and features.

The success of SMS is enabled by its innovative low-cost implementation concept which uses the signalling links of the GSM telephony system for the transport of short messages. The restriction to 160 characters for a short message and the relatively long transmission delays compared with many other data transmission services have not hindered the success of SMS. The high quality of the technical specifications and the test specifications and the test regime have ensured an end-to-end functioning including international roaming.

Another important success factor has been that the implementation of SMS was made mandatory for every GSM mobile and every GSM network by an agreement between the network operators in the framework of the GSM MoU Group. This led to an implementation in every mobile station and every network.

It then needed just a large population of terminals and the creativity of a lot of young people to make SMS a global success.

The additional features in the standard that were optional (such as EMS) were not so successful, as it was unclear for the message sender whether the recipient's mobile phone supported the option. In the case of EMS, for example, the sender was unaware whether the recipient's mobile could display the EMS message. Very few of these options achieved a significant market penetration, not least because no efforts were made to seek an agreement between network operators and manufacturers about the widespread implementation of the optional features.

9.2 Proposals for a Further Evolution of SMS: SMS Phase 3

The GSM market receives about 1 billion new terminals per year. In the second quarter of 2009, 286 million units were sold. Of these, 41 million were smart phones, which are capable of advanced messaging services.[1] Hence, a very large number of new SMS-capable phones will come on the market in the years to come. Therefore, it seems to be worth considering an evolution of SMS, including standardisation where needed. Areas for further evolution could be:

- A review of the specified optional features and an attempt to make those that improve the service mandatory. This could be done by a concerted effort on the part of operators and manufacturers. Standardisation is not needed.
- As SMS will for some time yet be used much more widely than Twitter, it may be worth implementing mechanisms that support user dialogues and the distribution of messages to several addresses (as in Twitter). However, as far as I am concerned, there is yet no protocol elements established between the SMS-SC and the MS to *manage* distribution lists stored in the SMS-SC. If that were catered for, SMS would take a great leap towards a more powerful chat platform. However, SC-specific short codes are already available for distribution to several addresses. Twitter has three modes of operation: message to all your followers, messages to specifically mentioned people and direct messages.

---

[1] Figures from Gartner press release, August 2009.

- SMS delivery receipts do not always work very well and are increasingly difficult as more and more gateways are included. This could be fixed with correct implementation and a push from the operators and/or manufacturers.
- Looking forward, there is a clear demand for SMS text messaging now and in the foreseeable future, and anyone developing new radio technology had better make sure that this market is catered for. This issue should be reviewed for LTE/SAE to ensure that precautions for SMS are made.

A workshop could be held that would develop a concept for an SMS Phase 3, and then a standardisation programme could be set up to specify it. In parallel, the buy-in of major operators and manufacturers should be sought to implement these features by a certain date. An SMS MoU could be signed to ensure success.

9.3 What Can be Learnt from SMS for Standardisation in Other Areas

9.3.1 General Aspects

Later standardisation has often not used the lessons learnt from SMS. These are the key lessons from SMS:

- Ensure that there is a value for the user and hence a market potential for a new feature.
- Keep the new feature simple. Hard limitations are acceptable if they lead to simplicity.
- Ensure a complete and robust specification and interoperability testing.
- Ensure the commitment of major network operators and manufacturers to implementing the new feature in terminals and networks from a certain time on.

9.3.2 A Proposal for Optimising MMS by a New, Simple, Ubiquitous MMS

MMS still has deficiencies owing to its complexity. The message container is structured like a slideshow. On each slide, many media types can be inserted. The originator of a message does not know the capabilities of the recipient. There are inconsistencies between different implementations. This version of MMS does not have the potential to become ubiquitous in every terminal.

However, there will be a very large number of new mobiles on the market in the years to come, as outlined at the beginning of Section 9.2. Therefore we would like to propose an additional new, basic, simple MMS suitable for ubiquitous use in every new terminal for the mass market. Here are the proposed key parameters of the basic MMS:

- *Improved signalling*: MMS uses a chargeable short message to set up the GPRS session ('establish a PDP context') for multimedia messages that need to be pushed towards a receiving phone, as such sessions cannot be set up by the network. This creates difficulties in IT for network operators and also in commercial considerations with regard to charging for the short message. In order to remedy this, SMS should be enhanced by introducing the possibility of using SMS messages as signalling elements that would not be shown to users and not charged like user messages.

- *Uniform message size for the multimedia message*: An implementation should be possible in every mobile. It is assumed that GPRS and EDGE will be available in every new mobile and in every network. In order to achieve reasonable download times, the message size should be limited to about 100 kB.
- *Multimedia Message format*: The message should be limited to one slide containing several media.
- *Photo*: The multimedia message may contain one photo in JPEG with a maximum size of about 50 kB. In email, the size of attached photos is normally reduced to 30–40 kB, and this is sufficient for the representation on all sorts of screen.
- *Text*: A field for about 1000 characters coded in 16-bit Unicode should be envisaged, which requires about 2 kB. This is roughly equivalent to the six SMS containers that are used in concatenated SMS. Such a size is also sufficient, as text inputting is a little difficult with the limited possibilities of a mobile station.
- *Speech/audio*: The remaining about 48 kB could be used either for about 60 seconds of speech[2] or for 10–15 seconds of audio[3]. This is sufficient to accompany a photo.

An agreement should be sought among major network operators and manufacturers that this feature package be implemented in every mobile and every network from an agreed date onwards. The commitment for mobiles should contain an obligation to be able to receive and to represent the full message. For sending, the omission of photos and audio should be possible, as this might be too expensive for low-cost mobiles.

This feature package is built according to SMS principles. It is very attractive to users. It provides a very great improvement on existing SMS and allows an implementation in every mobile. This feature would function reliably, as the sender knows the capabilities of the recipient.

---

[2] GSM half-rate codec assumed.
[3] MP3 or MPEG4 AAC.

Annex 1

Abbreviations Used in Several Parts of the Book

Abbreviation	Meaning
3G	Third Generation
3GPP	Third-Generation Partnership Project
AT commands	AT commands allow the terminal to control the modem. They were originally developed by the company Hayes Communications
BSC	Base Station Controller
BTS	Base Transceiver Station
CCITT	Comité Consultative International des Télégraphie et Téléphonie (later ITU T)
CEPT	Conférence Européenne des Administrations des Postes et des Télécommunications
CN	Core Network (TSG in 3GPP)
CPA	Content Provider Access
CR	Carriage Return (printer command)
CR	Change Request (to existing specifications)
CT	Core Network and Terminals (TSG in 3GPP)
DF900	Franco-German cooperation for GSM
DGMH	Drafting Group on Message Handling
EDGE	Enhanced Data Rates for the GSM Evolution
EMS	Enhanced Messaging Service
ERMES	European Radio Message System
ETSI	European Telecommunication Standards Institute
FMK	*Framtiden av Mobilkommunikation* (= The Future of Mobile Communications)
GMT	Greenwich Mean Time
GPRS	General Packet Radio Service

Short Message Service (SMS)   Edited by Friedhelm Hillebrand
© 2010 John Wiley & Sons, Ltd

Abbreviation	Meaning
GSM	(1) Group Spécial Mobile
	(2) Global System for Mobile Communication
GSM MoU	GSM Memorandum of Understanding
GSM#1, #2, #3, etc.	GSM Plenary Meeting #1, #2, #3, etc.
GSM1, 2, 3, 4, etc.	GSM Working Group or Subtechnical Committee 1, 2, 3, 4, etc.
GSMA	GSM Association
HLR	Home Location Register
IA5	International Alphabet No. 5
IDEG	Implementation of Data Services Experts Group
IMSI	International Mobile Subscriber Identity
ISDN	Integrated Services Digital Network
ISO	International Standardisation Organisation
ITU	International Telecommunication Union
LF	Line Feed (printer command)
MAP	Mobile Application Part
MHS	Message Handling System
MSC	Mobile Services Switching Centre
OMA	Open Mobile Alliance
OSI	Open Systems Interconnection
PAD	Packet Assembler/Disassembler
PCMCIA	Personal Computer Memory Card International Association, which gave its name to a standard for extension cards of mobile computers, known as PCMCIA-cards
PDU	Protocol Data Unit
PLMN	Public Land Mobile Network
PSTN	Public Switched Telephone Network
ROSE	Remote Operation Service Element
S900	Franco-German cooperation for an analogue 900 MHz system
SAPI	Service Access Point Identification
SIM	Subscriber Identity Module
SM	Short Message
SM-AL	Short Message Application Layer
SM-CB	Short Message Cell Broadcast
SME	Short Message Entity
SMG	Special Mobile Group (TC in ETSI)
SMG#1, #2, #3, etc.	TC SMG Plenary #1, #2, #3, etc.
SMG1, 2, 3, etc.	Subtechnical Committee 1, 2, 3, etc., to TC SMG
SM-LL	Short Message Lower Layer
SM-MO/PP	Short Message Mobile-Originated Point-to-Point
SM-MT/PP	Short Message Mobile-Terminated Point-to-Point
SM-RL	Short Message Relay Layer
SMS	Short Message Service
SMS-GMSC	Gateway MSC for the Short Message Service
SMS-IWMSC	Interworking MSC for the Short Message Service

Abbreviation	Meaning
SMS-SC	Service Centre for the Short Message Service (in the first WP4 documents also denoted 'SC')
SM-TL	Short Message Transfer Layer
STC	Subtechnical Committee (in ETSI)
T.50	Standard by ITU and ISO for alphabets to be used in telecommunication services
TC	Technical Committee (in ETSI)
TCAP	Transactions Capabilities Application Part
TP_AlertSC	Transfer Protocol – Alert SMS-SC
TP_MWD	Transfer Protocol – Messages Waiting Data
TP_MWF	Transfer Protocol – Messages Waiting Flag
TP_PID	Transfer Protocol – Protocol Identifier
TSG	Technical Specification Group (in 3GPP)
UMTS	Universal Mobile Telecommunication Service
VLR	Visitor Location Register
WP	Working Party (later called working group)
WP4	Working Party 4 Data Services
X.25	OSI protocol at the network layer (layer 3)
X.400	Standard by ITU and ISO for MHS

Annex 2

Sources for Quoted GSM Documents and Other Documents

The European Telecommunication Standards Institute (ETSI) maintains a comprehensive archive of most documents that have been created in the history of GSM: meeting reports including lists of participants, input documents, agreed documents, specifications of services and technical features. ETSI has published the documents of the Plenary of the GSM Committee in the DVD *GSM and SMG Archives 1983–2000*. This set of documents also covers the later SMS history fairly well. An updated version will be published in 2010, also covering early SMS history.

Then there is a description of GSM history including SMS in the book *GSM and UMTS: The Creation of Global Mobile Communication* (ed. by Friedhelm Hillebrand), published by John Wiley & Sons, Ltd, Chichester, UK, in 2001. This book presents contributions by the 37 people who participated in the development of GSM and UMTS. It touches upon the cornerstones of early SMS history and contains a CD ROM with key documents including those on the early history of SMS.

Where to find GSM Temporary documents?

Group	Format of document number	Used	Source(s)
GSM Plenary	GSM doc or TDoc 28/85	In GSM documents	• CD-ROM attached to the book *GSM and UMTS, The Creation of Global Mobile Communication*
	P-85-028	Filename on ETSI DVD and server	• ETSI DVD *GSM and SMG Archives 1983–2000*, updated version planned for late 2009 • ETSI Archive physical or: http://portal.etsi.org/docbox/zArchive/, then go to SMG and then to GSM groups

Short Message Service (SMS)   Edited by Friedhelm Hillebrand
© 2010 John Wiley & Sons, Ltd

Group	Format of document number	Used	Source(s)
GSM Working Party 1	WP1 37/85	In GSM documents	**Early documents before about 1994**
GSM Working Party 3	WP3 45/88		ETSI Archive physical or: http://portal.etsi.org/docbox/zArchive/, then go to SMG and then to GSM groups
IDEG	IDEG 05/85		
GSM Working Party 4	WP4 28/90		**Later documents after about 1994**
GSM4	GSM4 86/90		ETSI server: http://portal.etsi.org/docbox
SMG4	SMG4 22/94	In SMG documents	ETSI server: http://portal.etsi.org/docbox

Where to Find GSM Technical Specifications and Recommendations

Unfortunately there is no complete archive of GSM recommendations. They can be found in certain temporary documents. Here is a list of the key recommendations:

Recommendation	Version	Title	To be found in	Comment
GSM 02.03	Sept. 85	Teleservices supported by a GSM PLMN	WP1 12/85	First draft
	Nov. 86		IDEG 7/87 = WP1 70/86 rev. 2	Version approved by WP1
GSM 03.40	2.0.2 without annexes	Technical realisation of the Short Message Service Point-to-Point	P-88-212	First draft
	2.0.2 with annexes		P-89-045	Approved
	3.1.1		P-89-299	Approved
	3.2.2		WP4 63/83	Approved
	3.4.0		H&P archive	Approved
GSM 04.11	3.1.0	Support of SMS Point-to-Point on mobile radio interface	H&P archive	March 1990
GSM 09.02		MAP (support of SMS in network)	www.ETSI.org	

Where to Find Documents from the Franco-German Cooperation Programmes

Such documents should be in the archives of Orange France or T-Mobile Germany. A copy of many documents exists in the company archive of Hillebrand & Partners, Consulting Engineers.

Where to Find ETSI TC SMG Documents

See www.ETSI.org, access for ETSI members only.

Where to Find 3GPP Documents?

See www.3GPP.org, access for everybody.

Document Lists Used to Prepare Chapters 4, 5 and 6

The authors have used a fairly large number of documents to prepare these chapters. It is impractical to include lists of these documents in this book. They can be downloaded from http://www.gsm-history.org

Annex 3

Meetings of IDEG/WP4/GSM4 and DGMH in the Period from May 1987 to September 1990

The following meetings of IDEG/WP4/GSM4 and DGMH took place during the period from May 1987 to September 1990. Twenty-one meetings of DGMH were arranged during that period of time.

Meeting no./venue/month	Date of WP4 meeting	Date of DGMH meeting	DGMH Report (WP4/document number)
Meeting 1, Bonn, May 1987	20–22	–	–
Meeting 2, Heckfield, July 1987	6–8	–	–
Meeting 3, Helsinki, September 1987	8–11	8–11	71/87
Meeting 4, Bonn, October 1987	26–29	27–28	126/87
Meeting 5, Bonn, November 1987	23–27	24–26	151/87
Meeting 6, Oslo, January 1987	18–22	18–21	32/88
Meeting 7, Paris, February 1988	22–25	19–24[1]	87/88
Meeting 8, Oslo, March 1988	–	16–17	116/88
Meeting 9, Florence, April 1988	5–8	5–8	148/88
Meeting 10, Heckfield, May 1988	–	3–6	165/88
Meeting 11, Gothenburg, May–June 1988	30–3	31–2	208/88
Meeting 12, Luleaa, July 1988	4–8	4–8	251/88
Meeting 13, Oslo, September 1988	18–23	20–22	298/88
Meeting 14, Paris, November 1988	14–18	15–17	363/88
Meeting 15, Munich, January 1989	16–20	16–20	58/89
Meeting 16, Espoo, March 1989	–	14–15	65/89

Short Message Service (SMS)   Edited by Friedhelm Hillebrand
© 2010 John Wiley & Sons, Ltd

Meeting no./venue/month	Date of WP4 meeting	Date of DGMH meeting	Report (WP4/document number)
Meeting 17, London, April 1989	–	25–26	72/89
Meeting 18, Athens, May 1989	22–24	22–23	116/89 (List of CRs[2])
Meeting 19, Nottingham, September 1989	18–21	19–20	116/89
Meeting 20, Sophia Antipolis, November 1982	28–30	28–29	No report[3]
Meeting 21, Padua, February 1990	14–16	14–15	As above
Meeting 22, Bath, May 1990	9–11	9–10	As above
Meeting 23, Berlin, September 1990	10–14	11–13	128/90

[1] At this meeting, those attending DGMH had agreed, with the full support of the WP4 chair, to meet the Friday before the week of the WP4 meeting. In the period from 19 to 25 February, two DGMH meetings were conducted. However, in this survey they will be treated as one meeting.

[2] Only a table of Change Requests remains from DGMH in Athens (WP4 Document 116/89).

[3] At this meeting, the secretary of GSM4 prefers to incorporate reporting of the draft groups in the body text of the GSM4 MoM. However, in those cases, dates and those attending are omitted as well. Therefore, no separate report from the DGMH meeting was submitted.

Annex 4

DGMH Attendance in the Period from May 1987 to September 1990

Attending	Meeting no.																
	3	4	5	6	7	8	9	10	11	12	13	14	15	16	17	19	23
1. Eija Aaltonen[1]										x	x	x	x	x	x	x	x
2. Malcolm Appleby						x	x	x	x	x	x	x			x		
3. Jan Audestad					x												
4. Ermanno Berruto				x		x		x	x	x	x	x				x	
5. C. B. Dendé						x	x	x			x					x	
6. Trevor Callaway	x																
7. Robert Cohen											x	x	x			x	
8. Alan Cox			x			x	x			x	x	x	x	x		x	
9. Graham Crisp								x				x	x				
10. Nina Danielsen																x	
11. W. Fuhrmann			x														
12. Will Goodall							x										
13. Arthur Gidlow																	x
14. Michael Gießler														x			
15. Peter Gimskog	x	x															
16. Claudio Gentile	x																
17. Inge Groenbaek		x	x														
18. Jorma Haaranen	x													x			
19. Terje Henriksen		x	x	x	x	x	x	x	x								
20. Friedhelm Hillebrand						x											
21. Kevin Holley										x	x	x	x	x	x	x	x
22. Sverre Isaksen														x			
23. H.-E. von Istler	x																
24. Phil Knight							x	x	x	x							
25. Colin Farrell							x										

Short Message Service (SMS)   Edited by Friedhelm Hillebrand
© 2010 John Wiley & Sons, Ltd

Attending	Meeting no.																
	3	4	5	6	7	8	9	10	11	12	13	14	15	16	17	19	23
26. Bo Kvarnstroem				x	x	x	x	x	x	x	x						
27. Pekka Lahtinen														x	x		
28. Didier Luizard				x	x			x	x	x		x	x	x	x		
29. Pertti Lukander	x																
30. Ian Maxwell								x									
31. Francois Minet						x						x					
32. Simo Poikola	x				x												
33. Derek Richards																	x
34. Jon R Roernes										x	x						
35. David Rush					x	x		x									
36. Kauko Sallinen		x	x	x													
37. Lars Sandstroem														x	x		
38. Ruth Sharp															x		
39. Finn Skofteland										x	x	x	x	x	x		x
40. Paul Simmons									x				x				
41. Manfred Stahn					x				x								
42. Hans Thiger									x								
43. Finn Trosby	x	x	x	x	x	x	x	x	x	x	x	x	x	x	x	x	x
44. Aimo Vainio														x	x		
45. Knut Erik Walter						x					x						
46. Hans Wozny					x					x		x		x	x		

[1]Later, Eija Thiger.

Annex 5

Meetings of GSM4/SMG 4 and DGMH in the Period from October 1990 to the End of 1996

In the following, the meetings of GSM4/SMG4 and DGMH during the period from December 1990 to December 1996 are listed. Twenty-five meetings of DGMH were arranged during this period of time.

Meeting no./venue/month	Date of WP4 meeting	Date of DGMH meeting	Report (WP4/ document number)
Meeting 24, Paris, December 1990	3–7	5–6	001/91
Meeting 25, Ivalo, February–March 1991	25–1	26–28	067/91
Meeting 26, Stockholm, April 1991	16–18	–	080/91
Meeting 27, Vienna, May 1991	13–17	13–17	138/91
Meeting 28, Düsseldorf, September 1991	16–20	17–19	205/91
Meeting 29, Munich, October 1991	28–31	–	252/91
Meeting 30, Bonn, December 1991	2–6	3–5	001/92
Meeting 31, Paris, February 1992	24–28	24–27	079/92
Meeting 32, Firenze, May 1992	18–22	18–21	175/92
Meeting 33, Berlin, September 1992	7–11	7–10	253/92
Meeting 34, Brighton, November 1992	2–6	2–5	002/93
Meeting 35, Taastrup, February 1993	8–12	9–10	090/93
Meeting 36, York, May 1993	24–28	25–27	191/93
Meeting 37, Stockholm, September 1993	6–10	7–9	293/93
Meeting 38, Vienna, December 1993	6–10	8–9	001/04
Meeting 39, Stuttgart, March 1994	21–25	22–24	091/94
Meeting 40, Tampere, May 1994	16–19	17–18	165/94
Meeting 41, Cambridge, September 1994	13–15	13–14	251/94
Meeting 42, Sophia Antipolis, December 1994	13–15	13–15	003/95
Meeting 43, Düsseldorf, March 1995	6–9	7–9	080/95

Short Message Service (SMS)   Edited by Friedhelm Hillebrand
© 2010 John Wiley & Sons, Ltd

Meeting no./venue/month	Date of WP4 meeting	Date of DGMH meeting	Report (WP4/ document number)
Meeting 44, Uppsala, May–June 1995	30–2	30–1	159/95
Meeting 45, Turin, September 1995	25–27	25–27	283/95
Meeting 46, Sophia Antipolis, January 1996	8–11	8–11	156/96
Meeting 47, Lisbon, March 1996	26–29	26–29	304/96
Meeting 48, Paris, May 1996	20–23	20–23	420/96
Meeting 49, Tampere, September 1996	24–27	24–27	540/96
Meeting 50, Stratford-upon-Avon, December 96	9–13	9–13	698/96

Annex 6

DGMH Attendance in the Period from October 1990 to the End of 1996

There were more than 130 delegates in DGMH in the period. Therefore, the participation is split into two tables.

The report from meeting 34 is missing, so this column is blank. The two other blank columns are when there was no DGMH meeting.

Meetings 46, 47 and 49 had no list of DGMH delegates, so I have included all the people mentioned in the report – there is no other evidence of anyone attending the DGMH meeting (there is evidence of people attending the SMG4 meeting, but not all of them were in the DGMH meeting, and I think it unfair to include all the SMG4 delegates as DGMH participants).

Attending	Meeting no.																		
	24	25	26	27	28	29	30	31	32	33	34	35	36	37	38	39	40	41	42
Peter Albach					X		X	X	X	X									
Markus Aufmkolk													X	X					
Jacques Bellino																X			
Hossein Boroumand																		X	
Alan Carlton													X						
Alan Clapton													X						
Robert Cohen	X	X																	
Stephen Connor														X					
Louis Corrigan																		X	X
Francois Courau				X															
Paul Crichton																X			
Elizabeth Daniel																X	X	X	
John Davies																	X	X	

Short Message Service (SMS)   Edited by Friedhelm Hillebrand
© 2010 John Wiley & Sons, Ltd

Attending	Meeting no.																		
	24	25	26	27	28	29	30	31	32	33	34	35	36	37	38	39	40	41	42
Peter Decker												X	X	X					
Eric Desblancs																	X		
Eric Desorbay												X							X
Genevieve Devimeux												X	X		X		X	X	
Roland Edholm												X		X	X				
Peter Edlund	X	X										X	X			X		X	X
Heiko Eggers																X	X	X	X
Goran Eriksson														X	X	X	X	X	
Masoud Fatini							X	X	X										
Berndt Michael Fingerle																X			
Tim Fitzgerald												X							
Tom Forsyth				X															
Jeremy Fuller																			X
Arthur Gidlow	X	X		X	X		X	X	X	X			X	X		X	X	X	
Gerd-H. Groteluschen							X	X	X	X		X							
Ian Harris	X							X	X	X		X	X	X	X	X	X	X	X
Charlie Harrison				X															
Abdulla al Hashim																X			
Jurgen Hofmann													X						
Kevin Holley	X	X		X	X		X	X	X	X		X	X	X	X	X	X	X	X
Jukka Hosio																	X	X	
Chris Howard																		X	
Esa Ihamaki														X					
Jay Jayapalan	X	X																	
Timo Jokiaho														X					
Peter Krischan					X														
Anders Laage Kragh												X							
Robert Lambert																			X
Kim Lawson								X	X			X							
Frederic Leroudier							X	X	X	X			X	X		X			
Reinhard Lieberum										X		X	X	X	X		X		
Oscar Lopez-Torres										X									
Hans Jurgen Lorenz												X	X	X	X	X			
Jacques Lorentz												X					X		
Tommy Ljunggren								X											
Frank Mademann																X			
Annuziato Malara											X						X		
Dimitris Manolopoulos							X	X	X										
Timo Meuronen																X	X	X	X
Michel Mouly							X												
Lynn Mynors																		X	
Jose Nascimento																X	X	X	

Attending	Meeting no.																		
	24	25	26	27	28	29	30	31	32	33	34	35	36	37	38	39	40	41	42
Peter Neumann																		X	X
Jos Nooyen																X			
Mikko Palatsi							X												
Ian Park					X														
Wolfgang Patzak													X						
Bo Olsson												X							
Maria Papazoglou																X		X	
Malcolm Parrack																		X	X
Tibor Rako																			X
Markku Rautiola														X					
John Rees									X	X		X	X	X	X	X	X	X	X
Derek Richards	X	X		X	X		X	X	X	X				X					
Bernd Rieger												X							
Wolfgang Roth							X	X				X							
Donna Sand				X	X		X												
Ian Sayers										X		X							
Wolfgang Schott																X			
Thomas Schroeder			X					X	X			X							
Ante Sesardic															X				
Ole Christian Skredsvig								X	X				X						
Paul Simmons	X							X											
Finn Skofteland	X	X		X															
Lars Sorensen																X			
Eija Thiger	X	X		X	X														
Hans Thiger					X			X	X			X	X		X			X	X
Paolo di Tria												X	X						
Jonas Twingler									X										
Willy Verbestel									X										
Nisse Viklund						X													
Mike Watkins													X						
Keith Warren																		X	
Pauline Watt														X					
Mats Olof Winroth																			X
Hans Zschintzsch	X		X					X	X	X		X	X	X	X	X	X	X	X

The above table shows the attendance statistics of DGMH for meetings 24 to 42. Some participated only part time in certain DGMH meetings. Indications of this are not included in the table; for some meetings, no report exists and the column is blank; for a few meetings, no list of delegates exists but names are listed that are mentioned in the report.

Attending	Meeting no.							
	43	44	45	46	47	48	49	50
Peter Albach								
Markus Aufmkolk								
Chris Bailey								X
Nigel Barnes		X	X	X	X	X	X	X
Jurgen Baumann	X	X	X					
Jacques Bellino								
Craig Bishop								X
Celine Bluteau	X							
Hossein Boroumand								
Graham Brend								X
Christophe Cadoret		X						
Alan Carlton								
Christophe Choquet						X		X
Alan Clapton								
Robert Cohen								
Stephen Connor								
Louis Corrigan		X	X	X			X	
Francois Courau								
Paul Crichton								
Salvatore Cunsolo								
Elizabeth Daniel								
John Davies								
Peter Decker								
Eric Desblancs		X						
Eric Desorbay	X		X		X			
Genevieve Devimeux		X						
Koemar Dharamsingh						X		X
John Doyle							X	X
Pedro Duarte						X	X	X
Roland Edholm								
Peter Edlund	X	X	X		X	X	X	X
Heiko Eggers	X	X			X			
Goran Eriksson		X						
Don Erskine	X	X						
Masoud Fatini								
Denis Fauconnier		X				X		
Herbert Fedhalm								X
Berndt Michael Fingerle								
Tim Fitzgerald								
Tom Forsyth								
Jeremy Fuller	X	X						
Arthur Gidlow	X	X				X	X	X
Gerd-H Groteluschen								

Attending	Meeting no.							
	43	44	45	46	47	48	49	50
Patrik Gustaffson								X
Jari Hamalainen		X						
Ian Harris	X	X	X	X	X	X	X	X
Charlie Harrison								
Abdulla al Hashim								
Petri Heinonen			X			X	X	
Teri Hilteinen		X						
Jurgen Hofmann								
Kevin Holley	X	X	X	X	X		X	X
Jukka Hosio								
Chris Howard								
Andrew Howell	X							
Wolfgang Hultsch						X	X	X	X
Esa Ihamaki								
Jay Jayapalan								
Timo Jokiaho								
Peter Krischan								
Espen Kristensen		X						
Anders Laage Kragh								
Kimmo Laakkonen						X	X	X
Robert Lambert	X							
Kim Lawson								
Frederic Leroudier								
Reinhard Lieberum								
Janne Linkola								X
Oscar Lopez-Torres								
Hans Jurgen Lorenz								
Jacques Lorentz								
Tommy Ljunggren								
Frank Mademann	X							
Annuziato Malara								
Dimitris Manolopoulos								
Jose Martinez Zambrand								X
Massimo Mascoli	X							
Peter Mason								X
Timo Meuronen	X	X			X	X	X	X
Michel Mouly								
Lynn Mynors								
Jose Nascimento	X	X						
Tommy Ljunggren								
Frank Mademann	X							
Annuziato Malara								
Dimitris Manolopoulos								

Attending	Meeting no.							
	43	44	45	46	47	48	49	50
Jose Martinez Zambrand								X
Massimo Mascoli	X							
Peter Mason								X
Timo Meuronen	X	X			X	X	X	X
Michel Mouly								
Lynn Mynors								
Jose Nascimento	X	X						
Peter Neumann	X	X	X					X
Bernard Noisette								X
Jos Nooyen								
Reijo Nousiainen					X	X		
Lars Novak					X	X	X	X
Bo Olsson								
Mikko Palatsi								
Ian Park								
Wolfgang Patzak								
Maria Papazoglou								
Malcolm Parrack	X	X		X		X	X	
Isabell Perrot					X	X		X
Pauline Pike			X	X		X		
Graham Proudler		X						
Tibor Rako								
Markku Rautiola								
John Rees								
Derek Richards								
Bernd Rieger								
Ralf Roth								X
Wolfgang Roth		X			X	X		
Donna Sand								
Ian Sayers					X			
Andreas Schneeloch					X	X	X	X
Wolfgang Schott								
Thomas Schroeder								
Gavin Scott						X		
Ante Sesardic								
Finn Skofteland								
Ole Christian Skredsvig								
Paul Simmons								
Lars Sorensen								
Mr Steele								
Mr Tait					X			
Eija Thiger								
Hans Thiger								

Attending	Meeting no.							
	43	44	45	46	47	48	49	50
Philippe Thirion		X						
Paolo di Tria								
Joop Trouwee	X	X			X			
Jonas Twingler								
Willy Verbestel								
Nisse Viklund								
Mike Watkins					X	X		X
Keith Warren								
Pauline Watt								
Rhobert Weber	X							
Mats Olof Winroth						X	X	
Hans Zschintzsch	X		X					

The above table shows the attendance for meetings 43 to 50.

Annex 7

Evolution of GSM Specification 03.40

Meeting, date, location	Document	Content	Result
GSM#20, Oct. 88, Paris	P-88-212	03.40, version 2.0.2, without annexes	Noted
GSM#21, Jan. 89, Munich	P-89-045	03.40, version 2.0.2, including annexes	Approved
GSM#22, Mar. 89, Madrid	P-89-125	WP4 status: further study points identified for 03.40	
GSM#23, Jun. 89, Ronneby	P-89-223	Introducing default alphabet (taken from ERMES) and a coding scheme	
	P-89-225	Short message length set to 140 octets owing to MAP operation for inter-MSC transfer of short messages	Approved
	P-89-251	Removing variable interworking as a routing option	Approved
	P-89-252	Modifying Annex 1 owing to removal of variable interworking	Approved
	P-89-253	CR on lower-layer addressing	Approved
	P-89-254	CR on enabling the SMS-SC to distinguish between temporary and permanent errors	Approved

Short Message Service (SMS)   Edited by Friedhelm Hillebrand
© 2010 John Wiley & Sons, Ltd

Meeting, date, location	Document	Content	Result
	P-89-255	CR on defining certain values for the protocol identifier, e.g. for communicating with a variety of SME types	Approved
	P-89-279 (report comments)	Insertion of a mandatory alphabet, clarification of the protocol for the SMS-SC ↔ MSC connection (Annex 1) being optional	Approved
GSM#24, Oct. 89, Fribourg	P-89-299	Clarifying the update status for 03.40 version 3.1.1 (this version being attached)	Approved
GSM#25, Dec. 89, Rome	P-89-390	CR on use of originating address (MSISDN) for mobile-originated short message. A minor correction in the service definition of the transfer layer to avoid inconsistencies	Approved
	P-89-391	CR on multiple ISDNs in the alert in the Alert SC service element to align with GSM 09.02	Approved
	P-89-392	CR (editorial) to insert parameter diagram useful for introducing the flow of information included in a short message transferred between the SMS-SC and the MS	Approved
GSM#25bis, Jan. 90, The Hague	–	–	
GSM#26, Mar. 90, Sophia Antipolis	–	–	
GSM#27, Jun. 90, Stavanger	–	–	
GSM#28, Oct. 90, Corfu	P-90-334	Letter from PT12 informing that a minor error appeared in the alphabet schemes of 03.40 and 03.41. A recommended modification was enclosed	Noted
GSM#29 Jan. 91, Saar-brücken	P-91-003	Correct Greek characters, see Section 5.2.4 Correct ASN.1 in 03.40 Annex Clarify Validity Period bit order	Approved

Meeting, date, location	Document	Content	Result
GSM#30 Mar. 91, Bristol	–	–	
GSM#31, Jun. 91, Kiruna	–	–	
GSM#32, Sep/Oct. 91, Nice	P-91-309	SMS Reply Path, see Section 5.2.2	Approved
	P-91-310	SMS control characters	Approved
	P-91-313	Alphanumeric content in address field, see Section 5.2.11	Approved
SMG#01 Jan. 92, Lisbon	P-92-027	Address coding, message identifier	Approved
	P-92-061	SC-MSC protocol	Modified to P-92-101
	P-92-101	SC-MSC protocol	Approved
SMG#02 Mar./Apr. 92, Ostende	P-92-203	Alignment with MAP2 Operator-Determined Barring Memory Capacity Exceeded, see Section 5.2.8 Editorial Changes Status Report Codes Protocol Identifier Message Reference	Approved
SMG#03, Jun. 92, Copenhagen	P-92-307	SMS Command Multiple Message Transfer, see Section 5.2.1 Supplementary Services Memory Capacity Available Transfer Layer Parameters Replace Short Message, see Section 5.2.17 TP Message Reference Alert SC Clarification Annex Updates Default Values Editorial Changes	Approved, except Replace Short Message CR – modified in P-92-346
	P-92-325	07.05 first version, see Section 5.2.15	Noted
	P-92-346	Replace Short Message, see Section 5.2.17	Approved

Meeting, date, location	Document	Content	Result

SMG#04, Sept./Oct. 92, Madrid	P-92-413	Type of Number coding
		Message Classes
		Short Message Type 0, see
		    Section 5.2.17
		Address Length and Status Report
		Bit order for 7-bit alphanumeric
		    characters
		Additional Error Causes
		MS Controlled Replace SM
		R Interface
			Approved

SMG#04bis, Oct. 92, Paris	–	–	
SMG#05, Jan. 93, Amsterdam	P-93-016	Immediate Display PNE Rules Editorial SIM Specific SM, see Section 5.2.28 SMS Command Memory Capacity Available Notification TS-Status Report Optional/Mandatory	Approved
SMG#06, Mar/Apr 93, Reading	–	–	
SMG#06bis, May 93, Paris	–	–	
SMG#07, Jun. 93, Eindhoven	P-93-410	Correction of Errors SMS Interlayer Services	Approved

SMG#08, Sep./Oct. 93, Berlin	P-93-554	Check for SMS Duplicates, see
		    Section 5.2.22
		Return Call Message, see
		    Section 5.2.24
		Removing Service Primitives
		Service Centre Time Stamp, see
		    Section 5.2.25
			Approved

	P-93-558	SMS error causes	Approved
	P-93-585	Time zone coding, see Section 5.2.9	Revised in P-93-699
	P-93-699	Time zone coding, see Section 5.2.9	Approved

Meeting, date, location	Document	Content	Result
SMG#09 Jan. 94, Nice	P-94-042	Address field clarification	Approved
SMG#10, Apr. 94, Regensburg	P-94-243	Mapping of error causes	Approved
	P-94-244	Preparation for additional alphabets	Approved
	P-94-245	Clarification on temporary or permanent message failure	Approved
SMG#11, Jul. 94, Düsseldorf	P-94-350	Short Message Enquiry TS-Commands not specific to an individual SM Short Message Type 0 Enable Status Report Status Report Error Codes Status Report Reception Time	Approved
SMG#11bis, Sep. 94, Copenhagen	–	–	
SMG#12 Oct. 94, Helsinki	P-94-563	SM Duplicates, see Section 5.2.22 SM Congestion	Approved
SMG#13 Jan. 95, Sophia Antipolis	–	–	
SMG#14 Apr. 95, Rome	P-95-254	Concatenated SM Status Codes for SMS Commands Type of Number Coding Clarification of TP Message Number Problems with short messages having the same reference	Approved
SMG#15, Jul. 95, Heraklion	P-95-429	User Data Header SMS Command Responses PID for Internet Email, see Section 5.2.35 Editorial Corrections Concatenated SM, see Section 5.2.31 Problems with short messages having the same reference	Approved

Meeting, date, location	Document	Content	Result
SMG#15bis, Aug. 95, Paris	–	–	
SMG#16, Oct. 95, Vienna	P-95-607	Special message indicator, see Section 5.2.33	Approved
	P-95-608	3 CR editorial and clarification	Approved
SMG#17 Jan./Feb. 96, Edinburgh	P-96-068	Clarification and Editorial Changes	Approved
	P-96-073	UCS2 alphabet Liaison Statement from SMG4 to SMG and two CRs to 03.40 and 03.38 to introduce UCS2	Approved
SMG#18, Apr. 96, Bonn	P-96-242	SIM Data Download UCS2 Information Flow Diagram User Data Length MSISDN Alert in RP Error TP User Data Header padding	Approved
SMG#19, Jun. 96, Kista	P-96-394		
SMG#20, Oct. 96, Sophia Antipolis	P-96-607	ME de-personalisation over the air Internet Email, see Section 5.2.35 Unsupported values and upward compatibility	Approved
	P-96-608	Clarification Correcting TP-DCS RP User Data is optional in RP-Error over the radio Correcting table entries	Approved
SMG#21, Feb. 97, Paris	P-97-054	Enhanced diagnostic information for SM-MT, introduction of RP-ACK user data	Approved
	P-97-058	Introduction of SMS Compression for SM-PP and SM-CB service	Approved
	P-97-060	Additions of small features, editorial modifications or corrections	Approved
	P-97-180	Use of port numbers for interworking between SMS and higher-level applications	Approved

Meeting, date, location	Document	Content	Result
SMG#22, Jun. 97, Kristiansand	P-97-413	Technical enhancements and improvements	Approved
SMG#22bis, Aug. 97, London	–	–	–
SMG#23, Oct. 97, Budapest	P-97-696	SIM Toolkit Secure Messaging	Approved
	P-97-706	SMS transfer over GPRS	Approved
SMG#24, Dec. 97, Madrid	P-97-918	Secured SMS and SMS screening	Approved
SMG#24bis, Jan. 98, Paris	–	–	–
SMG#25, Mar. 98, Sophia Antipolis	P-98-0096	Allocation of a new code point for MS management	Approved
SMG#26, Jun. 98, Helsinki	–	–	–
SMG#27, Oct. 98, Prague	–	–	–
SMG#28, Feb. 99, Milan	–	–	–
SMG#28bis, Mar. 98, Frankfurt	–	–	–
SMG#29, Jun. 99, Miami	–	–	–
SMG#30, Nov. 99, Brighton	P-99-0675	Technical enhancements, improvements and error corrections	Approved
SMG#30bis, Dec. 99, Frankfurt	–	–	–
SMG#31, Feb. 00, Brussels	–	–	–
SMG#31Bis Apr. 00, Frankfurt	–	–	–
SMG#32, Jun. 00, Düsseldorf	–	–	–

Annex 8

Literature

- Friedhelm Hillebrand (editor, with contributions from 37 key players involved in the work for GSM and UMTS), *GSM and UMTS: The Creation of Global Mobile Communication*, John Wiley & Sons, Ltd, Chichester, UK, 2001, ISBN 0-470-84322-5.
- Gwenael Le Bodic, *Mobile Messaging Technologies and Services: SMS EMS and MMS*, 2nd edition, JohnWiley & Sons, Ltd, Chichester, UK, ISBN 0-470-84876-6.[1]

---

[1] http://eu.wiley.com/WileyCDA/WileyTitle/productCd-0470011432.html

Short Message Service (SMS)   Edited by Friedhelm Hillebrand
© 2010 John Wiley & Sons, Ltd

Annex 9

Brief Biographies of the Authors

Friedhelm Hillebrand

Friedhelm Hillebrand was born in Warstein, Germany, in 1940. He obtained a Master of Science degree in Communication Technology from the Technical University of Aachen in 1968. Then he joined the Research Institute of AEG Telefunken in Ulm and worked on system concepts for electronic switching systems. He joined Deutsche Bundespost, the national telecommunications carrier, in 1970. From 1976 on, he became responsible for the national X.25 packet switching network DATEX-P from the first concepts until the end of the first year of service in 1981.

He was then appointed project manager for the next-generation mobile communication system in spring 1983. He participated in GSM plenaries from November 1984 onwards. He contributed to services design, international roaming and work management methods. In 1987/88 he became the founding Chairman of the GSM data group (IDEG, later called WP4).

From mid-1987 he was responsible for the GSM contributions of Deutsche Bundespost, operators' cooperation in GSM MoU and the implementation of the D1 GSM network in Germany. He was responsible for the D1 system and network and the supporting IT systems until the end of the first year of operation in 1992.

Short Message Service (SMS)   Edited by Friedhelm Hillebrand
© 2010 John Wiley & Sons, Ltd

Fred was the first technical director of the GSM MoU Group from 1994 to 1996. There he worked on strategy and GSM promotion worldwide. From May 1996 to August 2000 he was chairman of the ETSI Technical Committee SMG. During this period the GSM Phase 2+ programme (e.g. GPRS, EDGE) and the fundaments of UMTS were completed. He proposed the creation of 3GPP.

Fred founded Hillebrand & Partners (H&P), Consulting Engineers, at the beginning of 2002. H&P have been specialising on IPR matters, such as analysis of essentiality of patents, search for prior art and the application of ETSI IPR policy, since 2004.

Finn Trosby

Finn Trosby was born in Sandefjord, Norway, in 1946. He was educated as a civil engineer at the Norwegian Technical University in Trondheim 1965–1970. After serving in the military forces (compulsory military service), he entered the research department in Telenor (at that time 'Televerket in Norway') in February 1972.

In 1980 he started to work on mobile communications, focusing on, for example, mobile messaging systems for both the private and business market segments. In 1987 he began work on GSM by taking part in the IDEG/WP4 group on data services for GSM. He was appointed chairman of DGMH (Draft Group on Message Handling, one of the subgroups of WP4) and remained in that post in the period from July 1987 to September 1990. DGMH was responsible for the design of the Short Message Service Point-to-Point and Cell Broadcast and for GSM access to the Message Handling Service (X.400) in fixed networks.

In 1990 he engaged in other R&D activities, e.g. the development of a simulator of traffic handling in a GSM network. In 1996 he entered Telenor Mobil – Telenor's mobile operating company in Norway, where he has been working on regulatory matters ever since. In latter years he has concentrated on issues related to spectrum and spectrum management.

Kevin Holley

Kevin Holley was born in Kingston-upon-Thames, UK, in 1963. He obtained his bachelor's degree in Physics at Bristol University 1982–1985. Following graduation, he joined the British Telecom Research Labs in Ipswich, UK, where he was responsible for various cellular radio work including the development a signalling monitor (hardware/software) for the analogue TACS system and developing a system including the user interface for an acknowledged radio paging technology (hardware/software).

In 1988 he started working on GSM, at first on a short project on call set-up times and then in layer 2 simulation work for data traffic, but moved quickly in June 1988 to work on SMS. He helped to complete the first SMS standard and started chairing the DGMH (Drafting Group on Message Handling) in September 1990. In 1997 he was appointed chair of the Special Mobile Group Working Group 4 (data services, including SMS) and he continued in that role through the change to 3GPP until mid-2001.

Since then he has been responsible for standards development in O2 and Telefónica, including chairing 3GPP TSG SA1, the OMA Requirements Working Group, vice-chair of OMA Technical Plenary and, since March 2009, vice-chair of 3GPP SA.

Ian Harris

Ian Harris was born at Minster on the Isle of Sheppey, Kent, UK, in October 1945.

On leaving Sheerness Technical School, he joined the Royal Air Force for a short term, specialising in air electronic radio and navigation, where he also studied electronic engineering at various colleges and universities. He obtained his higher national certificate in electronic engineering in 1966 and joined civil industry in 1968, pursuing a range of posts in R&D. These encompassed hardware and software design and development using the first 8-bit microprocessors in the mid-1970s for the data processing industry, through to the design and development of navigation and positioning systems for unmanned submersibles, ship and aircraft engaged in the offshore industry.

In 1984, he joined the newly created 'Vodafone' as their chief engineer for value-added services and was later appointed a senior technical manager. His primary role in Vodafone was for the specification, development and deployment of value-added services such as 'Data and Facsimile' over the UK analogue cellular total access communications system (TACS). With the advent of GSM, in 1988 he became Vodafone's design authority for SMS, working closely with his marketing peers to develop many business applications and enhancements for SMS until he left the company in 2002.

He obtained his Institute of Electrical Engineers (IEE) chartered engineer status in 1976, and was appointed a Fellow of IEE in 1991 for his work in Cellular Mobile Telecommunications.

He is currently a technical consultant to Research In Motion (RIM) – the company probably more well known for being the manufacturer of the 'Blackberry'.

He has been an active member in GSM and 3GPP standards since 1988 and was active in the creation of the first SMS specification. He was elected vice-chair of SMG4 (the ETSI group responsible for SMS) in 1998 and through its transition to T2 under 3GPP in 1999, when he took the responsibility of chairing the 3GPP T2 SMS subgroup. In May 2001, he was elected chairman of T2 and remained in that post until March 2005 when T2 was disbanded. He took on the task of rapporteur for SMS specifications in 1999 and retains that role today. He is still regularly consulted on SMS matters from across the industry.

Index

3GPP (Third-Generation Partnership
    Project) 43, 99, 106–9, 123–4, 141

abbreviations 135–7
addressing 62, 81–2
alphabets
    coding systems 92–3
    compression of new 97
    language tables 105–7
    see also character sets
application programming interfaces
    (APIs) 82–3
AT commands 94–5

bearer services 11–12
billing 122–3
binary messages, length of 81

character sets 54–5, 79
    expansion of 87–8
    for user information 64, 65
circuit-switched data services 3, 5, 11, 12,
    16
commercial applications 111
    billing 122–3
    CPA model deployed in Norway 123
    fixed-network connection to the
        SMS-sc 112–14
    intelligent terminal connections to
        mobile phones 116–17
    network operator interworking, roaming
        and number portability 114–15

performance 120–1
SMS in 2009 123–4
SMS keyboard text entry 117
SMS to fax and SMS to email 117–19
SMS traffic growth 121–2
third-party SMS-SCs 115–16
two-way real-time messaging
    application 119–20
compression 96, 97, 100–1
concatenated SMS 90–1
Content Provider Access (CPA) model,
    Norway 123
continuous message flow 77
control characters 90
Copenhagen Pizzeria session 17–18

delivery reports 78–9
DF900 *see* Franco-German cooperation
diversion of messages 91
documents 139–41
Drafting Group on Message Handling
    (DGMH)
    delegates 149–55
    first three years of 71–2
    meetings 143–4
        attendance (1987 to 1990) 145–6
        attendance (1990 to 1996) 147–8
    other tasks of 73–4
duplicates, avoiding 87

email, interworking SMS with 94, 117–18

Short Message Service (SMS) Edited by Friedhelm Hillebrand
© 2010 John Wiley & Sons, Ltd

enhanced messaging service (EMS) 101–3
error reporting, improvement of 86
European Radio Message System
    (ERMES) 54–5
evolution of SMS features
    (1990–96) 75–7
    Technical Improvements 77–97
evolution of SMS features
    (1997–2009) 99–100
    dissolution of T2 group 108–9
    enhanced messaging service
        (EMS) 101–3
    language tables 105–7
    other important standards work for
        SMS 107–8
    routers 104–5
    SIM toolkit data download and secure
        messaging 100
    SMS compression 100–1
    voicemail management 103–4
evolution of SMS, proposals for
    further 132–3
external systems, receiving SMS from 87
external terminal interface 84

facsimile (fax) technology 4–5, 11
    SMS to fax 117–18
fixed network services 2, 6
    bearer services 11–12
    reference model 9–10
    SMS-SC 112–14
    teleservices 10–11
Franco-German cooperation 19, 23–6
    significance of results 33–4
    standardisation proposal 32–3
    trial systems 26–7
    work on the SMS concept 28–31

General Packet Radio Service
    (GPRS) 107–8, 133–4
global market development
    acceptance of SMS 126–7
    infrastructure for 125–6

network operators' first use of
    SMS 126
    success of SMS 127–30
GSM (Global System for Mobile
    Communications)
    fixed-network-service
        companions 9–12
    GSM 04.11 72
    GSM4 meetings 143–4, 147–8
    memorandum of understanding
        (MoU) 43
    service philosophy (1982–4) 8–9
    specification, evolution of 157–63
    standardisation, methods of 7–8
    success of 12–13, 131–2

Harris, Ian 170–1
Hillebrand, Friedhelm 167–8
Holley, Kevin 169
Home Location Register (HLR) 58–9

icons for voicemail alert 90, 91–2
image encoding 102
Implementation of Data-Services Experts
    Group (IDEG)
    beginning of 43–4, 45–6
    instructions given for provision of
        SMS 47–8
    meetings 143–4
    see also WP4 (working party 4)
innovation vs. invention 15–16
Integrated Services Digital Network
    (ISDN) 16
intelligent terminal connections to mobile
    phones 116–17
interactive videotex 5
interface choice 68–9
international alphabet no. 5 (IA5) 30, 54,
    55
International Mobile Subscriber Identity
    (IMSI) 52, 60
international SMS messaging 97
interworking 69–70, 114–15

'invention' of SMS 15–18
    2009 discussion 18–19
    timetables of SMS development 19–21

keyboard text entry 117

language tables 105–7
length of short messages 54–5, 63–4,
    81–3
Long-Term Evolution (LTE) 124
long-term prospects 129–30

machine-to-machine communication
    (m2m) 130
MAP operations 54, 57, 61, 63, 72
market development *see* global market
    development
memorandum of understanding (MoU)
    43
memory capacity available 80
Message Handling Systems (MHSs) 6,
    73–4
message indication 91–2
Mobile Application Part (MAP) 54, 57,
    61, 63, 72
Mobile Services Switching Centre
    (MSC) 49, 58, 66–7
Mobitex 6
morse telegraphy 1
multimedia messaging service
    (MMS) 101–2, 133–4

National Group Number (NNG) 115
negative time zone 80–1
network architecture (1990) 57–60
network-based services (1980s) 1–6
network operators
    first use of SMS by 126
    interworking 114–15
networks, SMS between 94
Nokia cellular data card 86
non-numeric addresses 81–2
number portability 114–15

Open Systems Interconnection (OSI) 49,
    55, 57, 73

packet-switched services 3, 5, 6, 16, 66–7
performance 120–1
Phase 3 proposals for SMS 132–3
port numbers 107
predictive text 117
private mobile radio networks 6
Protocol Data Units (PDUs) 55–6, 82, 95,
    117
Public Land Mobile Network (PLMN) 38,
    58–60, 67–70

'R' interface to external equipment 84, 85
real-time messaging applications 119–20
Remote Operations Service Elements
    (ROSE) 49
Replace Short Message function 85–6,
    88–9
reserved code points 107
revenues 13, 127–9
roaming 94, 114
routers 104–5

S900 *see* Franco-German cooperation
server-based services 5–6
service centre *see* SMS-SC
Short Message Entity (SME) 58
shorthand text 117, 126
SIM *see* subscriber identity module
SMS concept creation (1984–87)
    acceleration of GSM project 42–4
    French and German network
        operators 23–34
    GSM committee and
        standardisation 34–42
SMS in 2009 123–4
SMS of September 1990
    addressing capabilities 62
    alphabet character set 64, 65
    maximum message length 63–4
    network architecture 57–9

SMS of September 1990 (*continued*)
   protocol architecture 60–1
   protocol design 55–7
   transfer schemes 61–2
SMS-SC (Service Centre for SMS) 47
   fixed-network connection to 112–14
   functionality of 104–5
   interconnect to cellular network 84–5
   location of 67–8
   multiple service centre scenarios
      77–8
   and network architecture 48–50,
      57–9
   service elements 50–3
   third-party 115–16
Special Mobile Group (SMG4) meetings
      (1990–96) 147–8
specifications, evolution of 157–63
standardisation 4–5, 11, 16, 19–20,
      107–8
   groups dealing with 43–4
   GSM way of working 7–8
   in GSM Committee (1985–87) 34–42
   lessons learnt from SMS 133–4
   proposal for 32–4
storage
   SMS in the phone 83
   Storing SMS on SIM card 80
Subscriber Identity Module (SIM) 83, 84,
   90
   storing SMS on SIM card 80
   toolkit/data download 100
success of SMS 12–13
   factors critical for 131–2
   global traffic and revenues 127–9

technical design (1987–90) 45–6
   description of work (1987–90) 48–55
   first three years 71–2
   instructions given to IDEG 47–8
   major design issues 64–70
   other tasks of DGMH 73–4

personal sentiments at the start 46–7
   SMS of September 1990 55–64
   work by other GSM bodies 72
technical improvements (1990–96)
      77–97
   from 1990 to 1991 77–83
   from 1992 84–6
   from 1993 86–9
   from 1994 89–91
   from 1995 91–6
   from 1996 97
telephony 2
teleservices 10–11
Teletex 5
Telex 2–3
terminal capabilities, detecting 86
time zones 80–1, 89
traffic in SMS, growth of 121–2, 127–8
Transaction Capabilities Application Part
      (TCAP) 63
transfer of SMS 48–53
   addressing capabilities 62
   inter-MSC 66–7
   layers 60–1
   schemes 61–2
Trosby, Finn 168–9
two-way real-time messaging 119–20

unacknowledged SMS 80
Unicode 93, 96, 105–7
Universal Mobile Telecommunications
      System (UMTS) 12–13, 107–8

Virtual Mobile Equipment (VME) 114
Visited Location Register (VLR) 58
Vodafone 111, 113–14, 117–18, 120–2,
   170
Vodafonem@il project 118–19
voicemail 88–9, 90, 91–2, 103–4

waiting messages, delay sending 89
WP1 (Working Party 1) 32, 35–42
WP3 (Working Party 3) 42–3

WP4 (Working Party 4)
   beginnings of 43–4
   meetings 143–4, 147–8
   technical design of SMS 45–55,
      69–70, 71, 73–4

X.25 protocol 49, 66–7
X.400 standard 73–4

youth culture 126–7

Printed and bound in the UK by
CPI Antony Rowe, Eastbourne

Printed and bound by CPI Group (UK) Ltd, Croydon, CR0 4YY

27/10/2024

14580292-0001